Intangible Natural Heritage

Routledge Studies in Heritage

1 **Intangible Natural Heritage**
New Perspectives on Natural Objects
Edited by Eric Dorfman

Intangible Natural Heritage

New Perspectives on Natural Objects

Edited by Eric Dorfman

Routledge
Taylor & Francis Group
NEW YORK LONDON

First published 2012
by Routledge
711 Third Avenue, New York, NY 10017

Simultaneously published in the UK
by Routledge
2 Park Square, Milton Park, Abingdon, Oxon OX14 4RN

Routledge is an imprint of the Taylor & Francis Group, an informa business

Typeset in Sabon by IBT Global.
Printed and bound in the United States of America on acid-free paper by IBT Global.

Library of Congress Cataloging-in-Publication Data
Intangible natural heritage : new perspectives on natural objects / edited by Eric Dorfman.
p. cm. — (Routledge studies in heritage ; 1)
Includes bibliographical references and index.
1. Natural history museums. 2. Natural history museums—Exhibitions. 3. Intangible property. 4. Cultural property—Protection. 5. Nature conservation. 6. Wildlife conservation. 7. Historic preservation. 8. Natural history. I. Dorfman, Eric.
QH70.A1I58 2011
508.075—dc22
2011015103

ISBN13: 978-0-415-88492-1 (hbk)
ISBN13: 978-0-203-13440-5 (ebk)

This book is dedicated to Dr Gerhard Winter, whose commitment in his former role as President of the International Council of Museums' Committee for Museums and Collections of Natural History (ICOM NATHIST), his continued support for its members and, in particular, his vision and initiative have catalyzed the creation of this book.

Contents

Figures

Foreword

This book, and the thinking around intangible natural heritage, has its roots, at least obliquely, in 2003 with UNESCO's seminal *Convention for the Safeguarding of Intangible Cultural Heritage*. The implicit assumption of the convention is that culture produces intangible heritage in the form of oral traditions, performing arts, social practices, rituals etc. Although UNESCO's definition includes "knowledge and practices concerning nature and the universe", thinkers in the field of natural history have felt that it falls short of providing a much-needed framework for acknowledging and fostering human relationships with the environment, not only for the benefit of society but for the ongoing conservation of nature. The fact that these relationships appear increasingly fragile makes meaningful discussion on the subject of intangible natural heritage especially timely.

Thus, when the International Council of Museums (ICOM) adopted "Museums and Intangible Heritage" for its 2004 general assembly in Seoul, the International Committee for Museums and Collections of Natural History (NATHIST) held a session of talks exploring approaches to intangible heritage in a natural history context. Following on from this, organizers of the 2006 annual NATHIST conference, in New Zealand, chose intangible natural heritage as the theme for papers at the gathering. Participants brought with them a rich and varied set of examples of their work, describing research, exhibitions and public programming. The present volume was catalyzed by both of these conferences.

Chapter 1, by Eric Dorfman, introduces the topic of intangible natural heritage and its philosophy, taking examples from fine art as well as natural history, and setting the stage for the diverse perspectives offered by the other authors. Inasmuch as tangible objects are the starting point for institutions thinking about intangible histories, the book begins with an exploration of the connection between them. Chapter 2, by Adrian Norris, provides a historical overview of the development of natural history collections, from the privately owned cabinets of curiosities to the more scientific collections of the present day. These collections have been subject to the dictates of fashion, as well as our understanding of science and biodiversity. The earliest of these collections were usually assembled for their appearance or novelty rather than scientific value. Modern, publicly -funded museums have developed in much the same way as the collections themselves, and Norris considers the impact of the vagaries of taste and changing professional practice the on today's collections, as well as ethical issues and the need for development of universal standards.

In the context of museums, people's ideas about nature most frequently begin with objects, extending beyond the physical to mix their own experiential knowledge, newly acquired information and innate responses to create novel understanding. Because much of the learning that happens in museums is self-paced and socially-mediated,[1] romance and emotion play a part, even in the context of science. In Chapter 3, John Long explores intangible natural heritage by focusing on a range of objects that are made valuable by their specific histories, rather than their material makeup alone.

How a natural history object can become a media phenomenon is the subject of Chapter 4, by Markman Ellis, who explores the reaction to a new species—the kangaroo—in the context of 18th-century Britain. Ellis explores the history of the animal's representations in London and in Sydney, focusing on three periods of collecting, display and representation of the kangaroo: the Cook expedition in the 1770s; the 'First Fleet' expedition lead by Captain John Phillips in Sydney Cove after 1788; and the commercial and scientific exploitation of kangaroo specimens in an unequalled 'media event' in London in the 1790s.

The many aspects of intangible natural heritage presented in this book are only useful if they add to programs, clarify roles, or in some other way benefit what museums are trying to accomplish. Chapter 5, by B. Venugopal, demonstrates how India's National Museum of Natural History incorporated intangible natural heritage into an important initiative called "Project INH", interpreting *Hortus Malabaricus*, a publication considered to be the earliest systematic documentation of traditional knowledge about natural heritage in Asia. *Hortus Malabaricus* is a unique publication, in 12 volumes published from the Netherlands by Van Rheede (who was earlier the Dutch Governor of Malabar) during 1678–1693. *Hortus Malabaricus* made extensive use of traditional knowledge of medicinal plants described from the "family books" of a few local experts, the originals of which are now lost. To promote this work in popular consciousness, Project INH consisted of a number of national conferences on *Hortus Malabaricus*, and catalyzed the establishment of a Museum Gallery on the topic.

Moving from the utility of plants for medicine to the value of landscape for recreation, Lee Davidson, in Chapter 6, considers the history of wilderness recreation in New Zealand and how it has come to encapsulate certain values in relation to natural heritage, e.g. 'natural quiet,' lack of human modification and an ethic of self-reliance or 'going on nature's terms.' Davidson then examines current threats to this traditional experience of nature—primarily from tourist development—and debates over the ways in which it might be preserved.

Intangible heritage is concerned in part with reaching outside the physical bounds of museums to focus on *in situ* processes and interactions. To do this, institutions must be aware of the values that visitors bring with them and, arguably just as important, how systems around those values develop. In the case of intangible natural heritage, this includes understanding community perspectives on wilderness and conservation. The ability of the traditional knowledge of nature to contribute to its protection is explored in Chapter 7, by Xingbao Jin and Alan Yen, who discuss the importance of crickets and katydids in Chinese culture. Both insects are vocal singers, widely distributed both in wilderness areas and urban green spaces. Keeping these insects is a hobby enjoyed by people from all walks of life, in a tradition that stretches back a thousand years. In addition to their melody, they are popular for their combative behavior, driving an active market for both wild-caught and captive-bred animals.

Thus, cultural diversity is transformed by natural diversity, and conservation of tangible natural heritage needs to be nourished by intangible cultural heritage. To answer this need, the Shanghai Science and Technology Museum uses singing insects as flagship species in concerts and exhibitions to inspire traditional culture for understanding the value of urban nature conservation. Their mission is to ensure protection of habitats for singing insects and, by extension, increase a sense of tranquility in urban Shanghai by promoting the insect sounds as a symbol for a society existing harmoniously with nature.

In Chapter 8, Ulrike Stottrop explores facets of meaning, in natural (both biological and nonbiological) objects and process. In it, she considers four exhibitions shown at the Ruhr Museum in Essen, and elsewhere in Germany, in the context of intangible natural heritage. Each provides a unique example of curatorial approaches which attempt to go "beyond the object". The exhibitions explore different methods of regional mapping, including olfactory memories, aural effects and the landscape of the blind person. They present objects, which compel the viewer to become aware of the natural processes that give the objects their final form, such as the impressions of raindrops from a 270-million- year-old shower.

The diversity of perspectives covered in this book underscores the nascent nature of the thinking around intangible natural heritage. As with any new field of endeavor, considerable thinking is needed in order to create the tools to put philosophy into practice. In Chapter 9, the final chapter of the book, Eric Dorfman and Janet Carding synthesize the ideas presented by the other authors in an attempt to bring together a cohesive framework for further discussion and research.

This project has been the brain-child of Dr Gerhard Winter, former President of ICOM NATHIST. His vision and initiative secured support for the project from the Paris-based head office of ICOM. The authors whose works appear in this volume were invited to contribute for the range of expertise and perspectives they represent, across a spectrum of the natural sciences, museology and visitor experience. Their efforts on behalf of this volume are greatly appreciated. Numerous people have helped with the production of this work. Clare Stringer, Curator Natural Science, Leeds Museum Discovery Centre provided the panda photo that appears in the Introduction, as well as information about the philosophy behind its display. We are grateful for the ongoing support of the President of ICOM, Dr Hanz-Martin Hinz, the President of ICOM NATHIST, Brigitta Schmid, and to the editorial staff at Routledge, Laura Stearns and Stacy Noto, who provided considerable support throughout the production process. We would also like to thank two anonymous reviewers, as well as Dr Andrew Barron, whose insightful comments much improved the work.

Eric Dorfman
Whanganui, New Zealand

NOTES

1. John Falk and Lynn D. Dierking, *Learning from Museums: Visitor Experiences and the Making of Meaning* (Walnut Creek: Alta Mira Press, 2000), 39.

1 Intangible Natural Heritage
An Introduction

Eric Dorfman

The state of the planet is declining—the climate is warming, deserts are encroaching on previously arable land, essential fish stocks are being depleted, and on every inhabited continent traditional societies are becoming subsumed by a Western lifestyle. As this happens, museums that house and interpret the world's natural and cultural artifacts are becoming increasingly involved with conserving the milieux from which their collections come. Museums are also becoming more active in raising public awareness about conservation, and are well placed to do so because they house the physical evidence of cultural and natural environments from both the past and present. The museum sector has begun to engage in an ethical dialogue around heritage conservation, which has resulted in the formation of the field of 'intangible heritage.' Since 2002, the subject has been gaining momentum in conversation, as a topic of research, and was the theme chosen for the International Council of Museums (ICOM) 2004 General Assembly in Seoul.

> *It is crucial to preserve our intangible heritage in order to perpetuate age-old traditions which may otherwise be lost forever and to place museum artifacts in their relevant context.*[1]

As reflected by this statement, conversation around intangible heritage has focused almost solely on cultural traditions, although the topic ostensibly includes both nature and culture. The philosophical underpinnings of intangible cultural heritage have begun to advance rapidly, aided by the United Nations Educational, Scientific and Cultural Organization's (UNESCO's) 2003 *Convention for the Safeguarding of Intangible Cultural Heritage.* It involves preserving the social traditions that work together to create the objects that eventually form museums' anthropological collections. These traditions comprise

> *the practices, representations, expressions, as well as the knowledge and skills, that communities, groups and, in some cases, individuals recognize as part of their cultural heritage.*[2]

Within this broad context, an enormous array of examples exists. For instance, the traditional *Girinya* dance of the Tiv people of Nigeria has been studied in light of changing cultural traditions, moving from a war dance that proscribed men's behavior in battle, to one invoking continuity and renewal.[3] Work is being done on the renaissance of whaling 'shanties' (occupation-related songs) at Barrouallie, a fishing community on the Caribbean island of Saint (St.) Vincent.[4] Additionally, the philosophy of contextualizing this research into practical museology is continually being developed.[5]

In the wake of increased interest in intangible cultural heritage, some natural history museums are beginning to turn their attention towards an environmental equivalent, 'intangible natural heritage.' Although still relatively new, the term is beginning to be used by some workers in museums and universities, especially exploring links to well-established fields of thought on natural heritage interpretation and conservation. One of its first uses in a public forum was in 2006, when it was the theme for the annual conference of ICOM's International Committee for Museums and Collections of Natural History (NATHIST).

This discussion comes at a time when museums are changing their focus. In 1992, Jean Davallon pointed out that museums were lagging behind in addressing environmental concerns:

> *Museums are not at the center of the powerful surge in interest shown by the public for environmental concerns, and this sets them far apart, might we add, from the press or television. Environmental issues are just barely creeping into the world of museums. Indeed, the literature on the subject of the environment –so abundant today –remains completely silent on museums and exhibitions. In the brotherhood of the mass media which constantly drives home the theme of the environment, museums go practically unnoticed. Only scientists, nature activists or visitors to exhibitions who are interested in the subject know that this theme is now part and parcel of modern exhibits. However, it must be said that museums rarely make the news on this topic and their style, with the exception perhaps of certain ecomuseums, hardly resembles that of the activist. In the final analysis, matters of substance seem to be taking place elsewhere, i.e., outside the museum. We might question, given these conditions, what is really at stake for museums in adding environmental themes to their exhibition agendas.*[6]

Almost 20 years later, this picture is at least beginning to change, although slowly, and arguably not to the extent that is needed. The most proactive natural history museums are those that are new, such as the Museums of Water and Life in Hemet, California, or which have been recently renovated, like the Paris Museum of Natural History, the Australian Museum and the California Academy of Sciences. Museums that are informed by best practice are taking a lead not only in telling the stories of natural heritage but also in innovating solutions to the challenges posed by operating sustainably. As these changes take place, the way that museums are thinking about the environment, as well as the stories they tell, are also broadening and becoming more complex. One palpable result of this is a move from the traditional chronological or taxonomic display of organisms to exhibitions based on themes such as traditional knowledge or conservation. Many museums are also actively engaged in outreach programs of education and environmental action, extending their influence beyond the confines of their walls. The *Birds in Backyards Project*, run jointly with Birds Australia and the Australian Museum is a good example of this.[7] Its aims are to increase public awareness and action around creating productive habitat for native birds in Australia's urban centers and to feed into ecological research to find out more about the needs of these bird populations. Work undertaken by the program's representatives includes online bird surveys, advice on creating bird-friendly spaces, an interactive bird finder and a public forum on urban birds and their habitats. Although never specifically stated, this program is sympathetic with the philosophies of intangible natural heritage.

Intangible natural heritage has, however, been used specifically by The National Museum of Natural History in India in describing its work conserving four species of hornbill[8] in north-eastern Indian state of Arunachal Pradesh. This area is a hotspot for both natural and cultural heritage, with five distinct ecosystems, as well as 26 major tribal communities holding diverse traditional knowledge about their biodiversity.[9] Economically, this region relies primarily on agriculture and rainforest products, which has had a major impact on both the extent and quality of habitat for hornbills and other wildlife.

Compounding this is the use of hornbill beaks by the men of some of the tribes to create headgear as part of a 'coming of age' ceremony. The impact is such that the species are highly threatened and the state's populations are now restricted to a single wildlife sanctuary. The museum has worked in partnership with the Wildlife Trust of India and the Forest Department of Arunachal Pradesh to create artificial beaks to use as substitutes in the costumes. They also worked with members of the Nyshis tribe through a lengthy trial and error period to gain acceptance for the beaks. Ultimately, they created a training program in which local people came to understand more about wildlife conservation and learned to make the beaks themselves, creating an economic incentive for preserving the species.[10]

The fact that this program was cited as an example of preserving intangible natural heritage raises the question of what, exactly, is encompassed by the term. Like 'biodiversity' or 'sustainability', intangible natural heritage might have many meanings, or shades of meaning, depending on the context and who is using it. If intangible natural heritage is to be a useful concept in establishing priorities and developing new ways of communicating with people both in and out of the museum, then it must be defined, at least for the people who are going to use it. This book is a first attempt at exploring some of the ways in which people are thinking about the term, focusing on examples and illustrating the way in which the concept fits alongside other fields of thought, such as conservation and interpretation. Ultimately, it aims to create a useful platform for further research and discussion.

INTANGIBLE NATURAL HERITAGE EXPLORED

Intangible *cultural* heritage is primarily concerned with what people do and how they express themselves within their social context. On one level, intangible *natural* heritage is an extension of this, encompassing indigenous knowledge of nature and conservation (see below). However, intangible natural heritage might also be considered to be far broader, because the forces that work to produce natural history objects are far more diverse, comprising both natural and anthropogenic forces. Thus, intangible natural heritage *might* encompass ethology, sociobiology and evolution (arguably analogous to the creative process that produces an object of cultural heritage), ecological food webs, landscapes, soundscapes, taphonomy, animism, opinions about wildlife in human societies, human food security and health, global climate, the conservation movement, taxidermy, bird watching and many others.[11]

A good illustration of the diverse ways in which intangible natural heritage could be approached is provided by a taxidermied giant panda *Ailuropoda melanoleuca*, on display at Leeds City Museum in the United Kingdom (UK; Figure 1.1). A member the bear family (Ursidae), the giant panda is native to central-western and southwestern China. Its striking black-and-white coat and outsized eye patches make it one of the most recognizable species on the planet. Its scientific interest is enhanced by its taxonomy, which has been debated for many years. Both giant and red pandas *Ailurus fulgens* were long thought to be members of the Family Procyonidae (raccoons and their relatives), and their taxonomy has been the result of a long-term and sometimes heated scientific debate.[12] Some of the reason for the disagreement is that, despite belonging to the order Carnivora, the giant panda is principally a vegetarian, having a diet that consists almost solely of about 25 species of bamboo. The history of this debate, and even the species' highly controversial use as an example of intelligent design,[13] contributes to the cultural fabric that gives the species meaning to us as members of society, whether we're visitors to museums or not. The fledgling topic of intangible natural heritage could encompass this history.

Visitors to Leeds City Museum will know about another topic that might fit well under the umbrella of intangible natural heritage: the species' conservation status. Giant pandas are critically endangered, their dependence on bamboo having made them extremely vulnerable to changes in

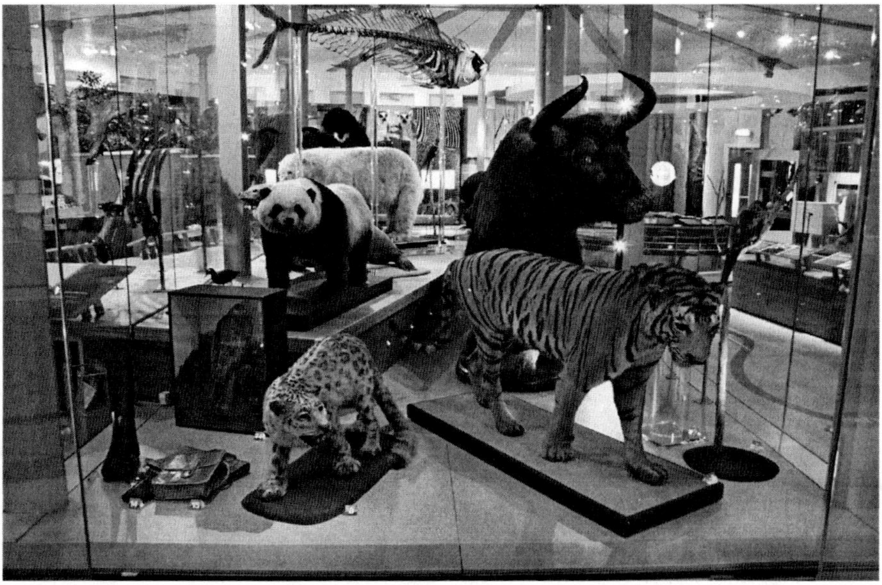

Figure 1.1 "Grandma" the giant panda *Ailuropoda melanoleuca*, displayed in the Leeds City Museum's science hall, in a multispecies display depicting "The Future of Biodiversity". Although the specimen is used by the museum to interpret conservation values, visitors will have many perspectives on pandas, which they will have formed before they ever encounter the display. Photo courtesy of Leeds City Museum, © Leeds Museums and Galleries, 2009.

land use. The species has been gradually forced from its former habitat—China's now heavily farmed lowland areas –becoming restricted to marginal alpine habitat in a few mountain ranges in the Sichuan, Shaanxi and Gansu Provinces of central China. Giant pandas' environmental issues are compounded by a natural bamboo die-off that occurs every 50 to 60 years, limiting habitat options further still. The species' position is made yet more precarious from poaching for fur and meat, and low birth rate due to a solitary lifestyle. Today, only between 1,500 and 3,000 giant pandas survive in the wild, and it is one of the few species in the world for which, in 2006, a UNESCO World Heritage Site has been designated.

Its plight, coupled with its attractiveness, has made the giant panda among the world's most adored and best protected species. When co-founding the World Wildlife Fund (now the Worldwide Fund for Nature, WWF) in 1961, Sir Peter Scott used the panda motif as its emblem. "Aware of the need for a strong, recognizable symbol that would overcome all language barriers, the [Founders of WWF] agreed that the big, furry animal with its appealing, black-patched eyes, would make an excellent logo".[14] Scott used as his model Chi-Chi, the giant panda residing at London Zoo. In so doing, he and his colleagues were aligning themselves to a public awareness phenomenon with considerable momentum.

Many adults visiting the Leeds City Museum would have seen Chi-Chi when they were children. She had come to the zoo in 1958, originally slated for a zoo in the U.S. However, this was during the Cold War, a time when America considered China "the strongest and most impregnable citadel of Stalinism",[15] and Washington had banned all trade. However, as she had already been collected, the Zoological Society of London took her, overlooking their previous conservation-based decision to discourage collecting wild pandas. Chi-Chi was immensely popular as soon as she arrived and failed attempts to mate her with Moscow Zoo's An-An always made front pages. When she died in 1972, Britain mourned her as one of the nation's best-loved animals.[16] Similarly, those too young to have seen Chi-Chi may well have watched a film of a sneezing baby panda on YouTube,[17] which, at the time of writing, has received nearly 10 million hits.

The high profile of giant pandas, coupled with the difficulty of captive husbandry and breeding, has catalyzed considerable research into both wild and captive foraging and mating behavior. The research is remarkable, both for its depth and breadth, in both wild[18] and captive[19] situations. Arguably few mammals, with the exceptions of bottlenose dolphins *Tursiops truncatus,* lions *Panthera leo* and the great apes are studied in greater detail. To the visitors at Leeds City Museum (and indeed of the entire developed world), these aspects of giant pandas combine to create a sense of familiarity—a background that they take with them into the exhibits. Although they might not know any of the specifics, these histories contribute palpably to their preconceptions.

Asian visitors to Leeds City Museum will bring with them a different set of preconceptions, as the popular appeal of giant pandas makes it a powerful marketing tool to Asian children. Products such as "Hello Panda" biscuits by the Singaporean bakery Meiji feature an anthropomorphic sport-playing cartoon panda on every fudge-filled cookie, and the pack features the same character waving at consumers, backed by a spray of bamboo leaves. Kabaya in Japan also make a panda-faced cookie. This product contributes funds to the WWF, and every pack features a page of cartoons, in Japanese, featuring cute antics of pandas in an urban setting. The threatened status of the animal also includes it, albeit obliquely, into the visual language of the global health food industry. American manufacturer Organic Foods uses the panda to advertise bars of peanut butter crisp, while they also appear on natural licorice chews produced in Finland. Perhaps least explicable are the many companies that use pandas in their name to market seemingly unrelated products such as beds (New Zealand), software (U.S.) and anti-slip adhesive floor tape (UK).

Possibly the most arcane aspect of this history is the way in which giant pandas contribute to cultural traditions of indigenous people. The hilly regions of the southern tip of Gansu Province of China are home to the Baima Tibetans, a unique ethnic group that still practices ancient religious customs founded in a rich traditional environmental knowledge. Their ideas about resource use, as well as their livelihoods, are connected to preserving local biodiversity, including that of the giant panda. It has been suggested[20] that this knowledge may be beneficial in conserving pandas, as well as overall regional biodiversity.

All of these diverse elements go to making the giant panda what it is as a physical entity and in relation to the human psyche. The only completely tangible aspects of the story are the animals themselves and, even then, individual specimens have histories that serve to connect them to the broader picture.

The specimen on display at Leeds City Museum was named "Grandma", and was one of 10 collected in Szechuan province for public display by the adventurer and professional animal collector Floyd Tangier Smith.[21] Despite her moniker, Grandma was only three years old at the time of death, having been named after the grandmother of his guide, who had her 80th birthday during the trip. Seven of the 10 animals collected were exhibited in Kowloon in 1938, three having died before reaching the exhibition. The surviving giant pandas were shipped to London, but a further two died on the voyage. The five remaining, Happy, Sung, Ming, Tang and Grandma, were sold in London on December 24, 1938. Happy was taken to the St Louis Zoo, and the others went to Regents Park Zoo in London. Grandma, however, died before the transaction was completed, after just 17 days in London. Depending on the account, her death was due either to double pneumonia or the amount of wood gnawed from her cage on board the *Antenor* in the final days of her voyage. There is an anecdote that Grandma was then passed to the Royal College of Surgeons who dissected the animal and wrote an illustrated report on the sexual organs of the female giant panda. It was then accessioned to the Leeds City Museum.[22] The remaining four pandas that arrived in 1938 resided at London Zoo and were featured in propaganda to boost morale during the WWII[23].

The tangible component of Grandma is her physical body, combined with the material used to stuff and mount her. The story behind Grandma links her to London Zoo and, indirectly, to Chi-Chi, as well as the history of conserving giant pandas. It also raises issues of captive animal welfare and the philosophy of collecting. All of this rich evolutionary, natural and social history combines to create the intangible components connected to the specimen.

Grandma is displayed in the Leeds City Museum's science hall, which opened in 2008. She is housed in a grouping of threatened animals, including a snow leopard and polar bear. The theme of the grouping is "the future of biodiversity"; the animals in the case are used to paint a bleak picture vanishing wildlife. They are tangible 'proof' of an intangible concept. They also, perhaps, serve to shock the visitor when they come face to face with a few of the victims of humanity's global spread. The purpose of engaging on intangible natural heritage is to inform conservation and interpretation. Sir

Peter Scott said, at the founding of the WWF, "We shan't save all we should like to, but we shall save a great deal more than if we had never tried".[24]

INDIGENOUS KNOWLEDGE

While intangible cultural heritage is primarily concerned with traditional understanding and expression that, in a museum context, contextualizes an artifact, indigenous knowledge is just one of myriad perspectives that can, arguably, be encompassed by intangible natural heritage. That said, indigenous viewpoints on conservation and natural process are important in creating a frame of reference for a great many natural objects. Just as the species themselves, this knowledge can often be threatened or even extinct.

Many mythological narratives and animals such as centaurs, cyclopes and giants, as well as a number of antique place names, have their origins in the enormous fossil bones of dinosaurs and large mammals that dominated lands later inhabited by the classical Greeks and Romans. For instance, griffins, the gold-guarding composite of eagle and lion, originated in stories from Scythian gold-miners, who encountered the skeletons of *Protoceratops* and other dinosaurs while passing through the Gobi Desert at the foot of the Altai Mountains.[25] This sort of information helps connect us to our past and allows for broader understanding of the importance and context of objects.

Indigenous knowledge can also provide additional insights into ecological process. In New Zealand, Māori traditional knowledge has identified an effect of climate change on *kina*, a sea urchin *Evechinus chloroticus*, used as a customary food. Shifting alignment between flowering of a local tree (*pohutokawa; Metrosideros excelsa*) and optimum *kina* harvest has been suggested to be an indication of changing climatic regimes.[26] Western science tends to discount qualitative observations such as these, because memory of past conditions can be distorted by subsequent events.[27] Nevertheless, these are valid as personal experiences and, in many cases, provide insight that can form the bases of testable hypotheses. Moreover, the people who made these observations are often intimately associated with the artifacts of interest to museums, making them a highly important part of the museum constituency.

Most native peoples have a concept of environmental stewardship wrapped into their social fabric. In Australia, it is told in stories of the Dreamtime,[28] to North America's Inuit, it is encompassed by "*Inuit Qaujimajatuqangit*" or "wisdom of the Inuit". The concept is both an epistemology and social program to incorporate traditional values into the practices of governance.[29] The Māori have a concept, *kaitiakitanga*, in which people (as well as spirits and other living creatures like trees) assist the gods (*kaitiaki*) in caring for the land. Responsibilities include restoring and maintaining *mana* (an impersonal force that inhabits and gives power to living things an inanimate objects) and *mauri* (life force) of the Māori, as well as protecting fragile ecosystems and ensuring plentiful resources.[30]

This is not to suggest that indigenous knowledge, by default, transcends Western conservation efforts. Indigenous communities are frequently just as culpable as the developed world of environmental degradation. Probably the most notorious example is the tragic mismanagement of Easter Island, sent from a fully forested island to a wasteland almost incapable of supporting life in little more than a 1,000 years of occupation.[31] The complex debate around traditional hunting and land use by Inuit in northern Canada and Alaska is another case in point.[32,33]

A sometimes heated controversy exists within sociology and conservation biology regarding whether community-based conservation is even possible. At one level, extractive resource use and economy, even a primitive one, is at odds with 'natural' ecosystems and processes,[34] and the strict conservationist view would be to depopulate sensitive areas completely. Despite the fact that this is rarely possible in practice, many thinkers are calling for a more holistic alternative that is both socially and ecologically sustainable.[35] The answer may come ultimately from a combination of cross-scale conservation, adaptive co-management, incentives and multiple stakeholders,[36] practices, which have proven highly effective, for instance in conserving wildlife in rural Africa.[37]

Institutions that collect, display and research natural history material have an ethical responsibility to engage with these issues.[38] This prescription was formalized by UNESCO in 2005:

> *Biological and cultural diversities are mutually reinforcing and interdependent. Natural systems cannot be understood, conserved and managed without recognizing the human cultures that shape them. Cultures cannot be divorced from their natural milieu that is the source of inspiration, creativity and sustenance.*[39]

Many museums have active wildlife conservation programs, and some of these incorporate indigenous knowledge into their activities. For example, the South African Museum in Cape Town aims to increase public awareness of indigenous knowledge though exhibitions on topics such as the concepts and skills of the hunter-gatherers of early Africa and the relationship between informal and scientific knowledge systems. Part of their interpretation involves the knowledge of medicinal plants held by traditional healers (still consulted by an estimated 70% of South Africans).[40] Countries, such as the U.S., Australia[41] and New Zealand,[42] which have institutionalized recognition of indigenous people, still struggle at a political level with incorporating traditional ecological knowledge into their legislation.

Irrespective of the ongoing debates, consideration of the issues enhances our understanding of ourselves and the world, informs the ethical display of collection objects and adds an important dimension to interpretation.

This broad base of representation is critical for museums to be of value to all of society, not just the educated, economically privileged echelon that produces the "typical" museum visitor.

INTERPRETING INTANGIBLE NATURAL HERITAGE

Developing a central unifying concept around intangible natural heritage is in line with developments in ideas of interpretation that have been maturing over recent decades.[43] Intangible natural heritage is concerned with processes, moving the viewer's attention beyond the object itself, to an abstract idea—a springboard for understanding process. As pointed out by Miles and Tout, moving on from this starting point is fundamental to visitors' understanding.

> *. . . the strongest argument against traditional object-and-label exhibits, in the context of the public understanding of science, is that the "failure to convey relevant concepts will prevent [the visitor] from achieving anything more than an illusion of understanding".*[44] *So it seems safe to suggest that objects in exhibitions do not, simply by being there, make possible a distinctive (or indeed any) understanding of science.*[45]

One of the last Great Auk *Pinguinus impennis* specimens in existence is stuffed and on display at the Senckenberg Museum in Frankfurt am Main, Germany. At one time this species was found in enormous numbers on islands off eastern North America as far south as Florida,[46] Greenland, Iceland, Norway, Ireland and Great Britain before being hunted to extinction.[47] Its demise is one of a litany of examples of the appalling environmental devastation by human society, especially since the Industrial Revolution. For those who know the story, it is a grim reminder of the destructive power of human beings. For those who don't, it is just another bird, which the uninitiated could be forgiven for mistaking for a penguin. Divorced from its history, the bird becomes an object of little intrinsic value (for the sake of argument, ignoring notions of nature's intrinsic value as expressed by Aldo Leopold and the deep ecology movement). It is the intangible component of its natural heritage that gives it worth, both intellectually and monetarily.

At the other end of the same spectrum, one of the most intrinsically valuable objects ever created is Damian Hirst's sculpture *For the Love of God*. It consists of a platinum cast of a human skull encrusted with 8,601 flawless diamonds, including a large pear-shaped pink diamond located in its forehead. It cost £14 million to produce and, when the work went on display at in London, it had an asking price of £50 million. Had it sold, it would have brought the highest price ever paid for a single work by a living artist. It has been displayed with gravitas, in a darkened room, by which visitors pass almost in single file, many commenting on its enormous monetary value.

On one level, it could be considered a natural history object (as are many of Hirst's other works). It is based on a common human skull, created out of metal and stones with real teeth. However, in contrast to the Great Auk (Figure 1.2), a great deal of the value of *For the Love of God* lies in the materials from which it was made. The work does not need background information to validate or explain it to those who view it; the images are those of common experience. Only its worth is phenomenal. In fact, even the idea of covering a skull with jewels is not new. Artist John LeKay, a friend of Hirst's in the early 1990s, claims that *For the Love of God* is based on *Spiritus Callidus #2* a similar skull covered with crystals, which LeKay had made in 1993. One suspects, however, LeKay's work in an inexpensive material would have remained obscure had it not been for its connection to Hirst's. Hirst's work transcends the purely visual by the irresistible juxtaposition of fabulous wealth with a literal symbol of death. Like the Auk, it was what we know (or can learn) about the object that gives it deeper meaning.

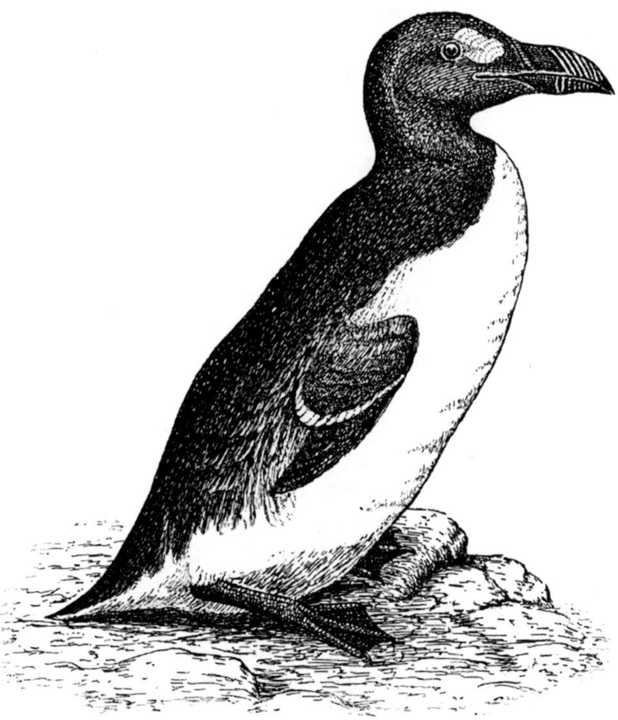

Figure 1.2 The Great Auk *Alca impennis* (=*Pinguinus*), from a woodcut by Edward Forbush. In 1912, when his work came out, the Great Auk was already extinct. Edward H. Forbush, *Game Birds, Wild-Fowl and Shore Birds of Massachusetts and Adjacent States* (Boston: Massachusetts State Board Agriculture, 1912).

An exhibition on malaria would present a completely different challenge, that of displaying an issue almost devoid of objects. Malaria is a blood disease caused by the parasite *Plasmodium* and carried by the *Anopheles* mosquito. The disease currently afflicts more than 250 million people, causing close to a million deaths every year, and puts about half of the world's population at risk.[48] It has been important historically, affecting or killing many great people, including Alexander the Great, Dante, a number of popes, Lord Byron, various U.S. Presidents from Washington to Kennedy, Mahatma Gandhi, Ernest Hemmingway and Mother Theresa.

National and international authorities wage an ongoing war in an attempt to control malaria and counteract its effects. Methods range from administering drugs to controlling mosquitoes and issuing those at risk with mosquito nets. Dichloro-Diphenyl-Trichloroethane (DDT) is one of the most effective insecticides and is used extensively, despite having been banned in many countries because of its effect on birdlife.[49].

Human behavior has a role in the spread of malaria, so public awareness is vital.[50] This need occasionally results in education-based exhibitions, such as one held in Beijing in 2006 as part of the Forum on China-Africa Cooperation. Choices for objects in these exhibitions are limited and tend to be small and, in themselves, not very impressive. Mosquitoes under microscopes, coupled with photographs of the stricken, micrographs, maps, pie charts and life-cycle diagrams are the most common. Despite being one of the world's most important diseases, it is an interpretation challenge that has, thus far, largely evaded the museum community.

The World Health Organization has for decades used postage stamps as a means of raising awareness of malaria, of prevention and of commemoration. Almost every nation has participated at some time and, in doing so, has given momentum to a trend of displaying this disease (and other current issues) on stamps as awareness-raising programs. It has become an important activity for some collectors, as evidenced by Malaria Philatelists International,[51] an association that since 1979 has been devoted to collecting and archiving malaria-themed stamps.

From the standpoint of interpreting intangible natural heritage, creating, collecting and displaying stamp issues is relevant because the activity has the power to transmogrify oblique or even inscrutable ideas into common objects used to carry out the mundane task of posting a letter. In a sense, stamp issues are exhibitions in miniature; instead of filling the halls of an institution, they are brought into the home as an accessible, utilitarian item. If this interpretation is successful, those who purchase stamps, or receive them in the post, will perceive the pictures as tiny windows into an abstract idea.

However we chose to do it, interpreting, collecting and conserving intangible natural heritage requires a cogent understanding of what, exactly, it is. In doing so, we are forced to re-examine our relationship with nature. However, developing this understanding is not easy, partly because the topic is so new, and partly because it is potentially so broad. The remarkably different approaches to thinking about interpretation and museology form the basis of this book. The chapters that follow are the first attempt to synthesize different approaches to intangible natural heritage and represent a cross-section of the broad range of current thinking.

NOTES

1. ICOM, "Seoul 2004", *ICOM News* 57, no. 4 (2004): 3.
2. UNESCO, *Convention for the Safeguarding of the Intangible Cultural Heritage*, 2003.
3. James M. Agaku, "The *Girinya* dance theatre of the Tiv people of Nigeria: an aesthetic evaluation", *Journal of Intangible Heritage* 3 (2008): 30–41.
4. Daniel Lanier and Vincent Reid, "Whalers' shanties of Barrouallie, St. Vincent: observations on the nature, decline and revival of a unique Caribbean maritime tradition", *Journal of Intangible Heritage* 2 (2007): 70–80.
5. See, for instance, Marilena Alivizatou, "Contextualizing intangible cultural heritage in heritage studies and museology", *Journal of Intangible Heritage* 3 (2008): 44–54.
6. Jean Davallon, Gérald Grandmont and Bernard Schiele, *The Rise of Environmentalism in Museums/L'Environnement entre au musée* (Lyon: Presses Universitaires de Lyon, 1992), 206.
7. Eric Dorfman, Annabel Riley and Kate Ravich, "Birdscaping the Suburbs: The Australian 'Birds in Backyards' Programme and Its Potential for Use in Wellington", in *Greening the City: Bringing Back Biodiversity Back to the Urban Environment,* ed. Murray Dawson (Christchurch: Royal New Zealand Institute of Horticulture, 2005), 136–144.
8. Great Hornbills *Buceros bicornis,* Wreathed Hornbills *Rhyticeros undulatus*, Oriental Pied Hornbill *Anthracoceros albirostris* and Rufous-necked Hornbill *Aceros nipalensis.*
9. Ranjay H. Singh, "Biocultural knowledge and Adi community: conservation and sustainability in biodiversity hotspot of Arunachal Pradesh", *Current Science* 96 (2009): 883–884.
10. Sushil K. Saraswat, "Natural History Museums on the Turf for the Challenging Roles to Play for Conservation of Intangible Heritage—An Indian Experience" (paper presented at the ICOM International Committee for Museums and Collections of Natural History, Seoul, South Korea, October 2–8, 2004).
11. In that humans are biological entities, intangible natural heritage could be seen to encompass intangible cultural heritage as well. However, for clarity, the two are left distinct.
12. See, for example, Stephen J. O'Brien, William Nash, David Wildt, Mitchell Bush and Raoul Benvensite, "A molecular solution to the riddle of the giant panda's phylogeny", *Nature* 317 (1985): 140–144.
13. Percival Davis and Dean Kenyon, *Of Pandas and People: The Central Question of Biological Origins* (Richardson: Foundation for Thought and Ethics, 1989).
14. "What Is the Story behind the Logo of the WWF?" accessed February 22, 2011, http://wwf.panda.org/what_we_do/endangered_species/giant_panda/panda/panda_evolutionary_history/.
15. Ygael Gluckstein, *Mao's China* (Boston: The Beacon Press, 1957), 421.
16. Michael R. Brambell, "London Zoo's Giant Panda: *Ailuropoda melatiolenca*: 'Chi Chi,' 1957–1972", *International Zoo Yearbook* 14 (1985): 163–164.

17. Jim V. W. Moss, "Sneezing Panda", accessed February 22, 2011, http://www.youtube.com/watch?v=FzRH3iTQPrk.
18. See, for instance, Xue Hua Liu, Andrew K. Skidmore, Tie Jun Wang, Yong, Yange and Herbert H. T. Prins, "Giant panda movements in Foping Nature Reserve, China", *Journal of Wildlife Management* 66 (2002): 1179–1188.
19. See, for instance, Mollie A. Bloomsmith, Marvin L. Jones, Rebecca J. Snyder, Rebecca A. Singer, Wendy A. Gardner, Ivy S. C. Liu, and Terry L. Maple, "Positive reinforcement training to elicit voluntary movement of two giant pandas throughout their enclosure", *Zoo Biology* 22 (2003): 323–334.
20. Yaofeng Luo, Jinlong Liu and Dahong Zhang, "Role of traditional beliefs of Baima Tibetans in biodiversity conservation in China", *Forest Ecology and Management* 257 (2009): 1995–2001.
21. A fascinating side story exists concerning his rival Ruth Harkness, the socialite turned adventurer (and widow of Smith's former business partner) who, in 1939, beat Smith in obtaining the first panda for an American zoo. See Vicki Croke, *The Lady and the Panda* (New York: Random House, 2005).
22. Adrian Norris, "Notes on a giant panda", *Leeds Naturalists' Newsletter* 3 (1981): 43–46.
23. Jon Clarke (ed.), "The Good Zoo Guide: London Zoo", accessed February 22, 2011, last modified February 22, 2011, http://www.goodzoos.com/UK%20Zoos/london.htm.
24. "What Is the Story behind the Logo of the WWF?"
25. Adrienne Mayor, *The First Fossil Hunters: Paleontology in Greek and Roman Times* (Princeton: Princeton University Press, 2001), 28.
26. Alan Parker, Zoltán Grossman, Edward Whitesell, Brett Stephenson, Terry Williams, Preston Hardison, Laural Ballew, Brad Burnham, Jill Bushnell and Renée Klosterman, *Climate Change and Pacific Rim Indigenous Nations* (Olympia: Northwest Indian Applied Research Institute, 2006), 29–38.
27. This can also be manipulated by advertising. See, for instance, Jianqin Yang and Robert Meeds, "Effects of false post experience advertising, message processing environment, and nationality on memories for personal experiences", accessed February 22, 2011, *All Academic Research* 172883 (2008): 1–29, http://www.allacademic.com/meta/p172883_index.html.
28. See, for instance, Jim Poulter, *The Secret of Dreaming: The Story of Why the Land Is Sacred and Why Man Must Be Its Caretaker* (Templestowe: Red Hen, 1988).
29. Pamela Stern, *Historical Dictionary of the Inuit* (Lanham: Scarecrow Press, 2004), 72.
30. Andrée Mathieu, "Kaitiakitanga –Safe Guarding Our Future", accessed February 22, 2011, http://www.kaitiakitanga.net/stories/origins%20research.htm.
31. Clive Ponting, *A Green History of the World* (London: *Penguin Books*, London, 1992).
32. Colin Irwin, "Lords of the Arctic: wards of the state", *Northern Perspectives* 17 (1989), accessed February 22, 2011, http://www.carc.org/pubs/v17no1/2.htm.
33. Tungavik Federation of Nunavut, "An Inuit response", *Northern Perspectives* 17 (1989), accessed February 22, 2011, http://www.carc.org/pubs/v17no1/4.htm.
34. Janis B. Alcorn, "Indigenous Peoples and Conservation", *Conservation Biology* 7 (1993): 424–426.
35. See, for instance, Marcus Colchester, *Salvaging Nature: Indigenous Peoples, Protected Areas and Biodiversity Conservation*. United Nations Research

Institute for Social Development Discussion Paper 55 (1994), accessed February 22, 2011, http://www.wrm.org.uy/subjects/PA/texten.pdf.

36. Fikret Berkes, "Rethinking community-based conservation", *Conservation Biology* 18 (2004): 621–630.
37. David Hulme and Marshall Murphee (eds.), *African Wildlife and Livelihoods: The Promise and Performance of Community Conservation* (Oxford: James Currey, 2001).
38. The morality of indigenous ecological knowledge has been addressed by Fikret Berkes, "Traditional Ecological Knowledge in Perspective", in *Traditional Ecological Knowledge, Concepts and Cases*, ed. Julian T. Inglis (Ottawa: Canadian Museum of Nature, 1993), 1–9.
39. UNESCO, "Safeguarding the Transmission of Local & Indigenous Knowledge of Nature", Draft Working Document. *Exploring Linkages between Cultural Diversity and Biological Diversity* (Paris: UNESCO, 2005), accessed February 23, 2011, http://portal.unesco.org/science/fr/ev.php-URL_ID=4489&URL_DO=DO_TOPIC&URL_SECTION=201.html.
40. Anthony B. Cunningham, "African medicinal plants: setting priorities at the interface between conservation and primary healthcare". *People and Plants Working Paper 1* (Paris: UNESCO, 1993), accessed February 22, 2011, http://unesdoc.unesco.org/images/0009/000967/096707e.pdf.
41. Anne Ross and Kathleen Pickering, "The politics of reintegrating Australian Aboriginal and American Indian indigenous knowledge into resource management: the dynamics of resource appropriation and cultural revival", *Human Ecology* 30 (2002): 187–214.
42. "New Zealand's Controversial Foreshore and Seabed Act (2004) Has Been Called 'the Most Significant Piece of Legislation Affecting Maori in Recent History' by the Ngai Tahu Tribe", accessed February 23, 2011, http://www.ngaitahu.iwi.nz/Te-Runanga/He-Kaupapa-Whakahirahira/Foreshore-Seabed/.
43. See overview by Graham Black, *The Engaging Museum: Developing Museums for Visitor Involvement* (London: Routledge, 2006).
44. Brian N. Lewis, "The museum as an educational facility", *Museums Journal* 80 (1980): 151–155.
45. Roger Miles and Alan Tout, "Exhibitions and the public understanding of science", in *Museums and the Public Understanding of Science*, ed. John Durant (London: NMSI, 1992), 26–27.
46. Penelope Hermes Weigel, "Great Auk remains from a Florida shell midden", *Auk* 75 (1958): 215–216.
47. BirdLife International, "Pinguinus impennis", in *2006 International Union for the Conservation of Nature and Natural Resources (IUCN) Red List of Threatened Species* (Paris: IUCN, 2006).
48. World Health Organization (WHO), *World Malaria Report 2008* (Geneva: WHO, 2008), viii, accessed February 21, 2011, http://whqlibdoc.who.int/publications/2008/9789241563697_eng.pdf.
49. Rachel Carson's seminal book *Silent Sprint* was responsible for the ban of the product in the U.S. Rachel Carson, *Silent Spring* (New York: Houghton Mifflin, 1962).
50. For instance, in many African countries, the proportion of family members (children and pregnant women) that slept under an insecticide-treated mosquito net is typically smaller than the proportion of households that own one. WHO, *World Malaria Report 2008*.
51. "Malaria Philatelists International (MPI)", accessed February 22, 2011, http://www.malariastamps.com/mpi/index.asp.

2 The Intangible Roots of Our Tangible Heritage

Adrian Norris

INTRODUCTION

Communicating the intangible as well as presenting the tangible is part of the work of museums. Early explorers, writers, scientists and collectors all in their own way tried to explain the natural world to their peers. Mythology accompanied artifacts brought back from exotic lands, with mysterious objects being displayed in cabinets of curiosities. These early forerunners grew into the global family of museums to which so much of our heritage is entrusted. A large part of the museums' reason for being was, and is, to exhibit and explain to a culturally diverse audience the intangible aspects of our planet. However as they developed, museums did not always fulfill that purpose. Poor collecting and documentation practices, as well as a lack of understanding of other cultures' beliefs and knowledge, sometimes placed barriers between peoples' differing cultural values. Financial constraints restricted how these growing collections developed and rising costs sometimes resulted in the destruction and dispersal of collected material without reference to its worldwide cultural value.

Museums have evolved into global institutions from which new and important advances can emerge. If they are to develop further their rightful place in the global scientific community they need to do more to ensure that information held in, and on, collections, both tangible and intangible, is available to all interested parties: environmentalists, conservationists, educators, researchers and even armchair naturalists through such means as the Internet. New scientifically advanced fields, such as research into the possibilities of cloning extinct species, may also help to bring new life and resources into safeguarding and developing the natural science collections held in our museums.

However, despite its potential scientific utility, an exotic animal held in any specific museum is also a part of the fauna of its country of origin and remains part of the culture of that country in the same way as a piece of national costume or an archaeological artifact.

In this chapter, I review the growth of museums from their earliest forms, in the context of peoples' curiosity about nature and demonstrate how museological practice in the past has stripped many natural historical objects of their intangible aspects, because museums (both nascent and actual) were shaped by external forces that valued only the tangible. I will discuss how intangible heritage, if it was collected at all, was easily lost thereafter, and how contemporary practice is not immune to this trend, with the state of taxidermy presented as an example.

MYTHICAL AND EXOTIC WORLDS

The origins of natural history collecting predate our ability to record it. Since time immemorial, humans have been fascinated and sometimes terrified by the alien creatures found outside their immediate circle of experience. Roman legionaries brought home objects as tokens or mementoes of their experiences in foreign lands. One legionnaire is known to have made a collection of seashells while in the area of the North Sea, transporting them on foot over what at the time was a vast distance back to Rome. Alexander the Great returned from conquest with exotic faunas and floras of distant lands in much the same way as explorers of later centuries.[1]

Mediaeval soldiers, sailors, explorers and missionaries all brought back to their respective countries tales of monstrous beasts and birds that inhabited the areas of "*Terra Incognita*" and "*Mare Incognito*". Tales of strange cultures and even stranger animals caught the imagination of thinkers throughout the known world. In the Middle Ages, serious theological debates took place in Europe as to the origin of such creatures, some depicted as being half-human and half-animal, and whether they had souls and should thus be converted to Christianity or not, if and when encountered by missionaries. John Hale wrote "*An age of intellectual ferment, as the Renaissance undoubtedly was, is not necessarily open-minded*".[2] Explorers and travelers were expected to find monsters, particularly in Asia, The Americas and Africa (the last two thought to be a single continent).

In the early Renaissance, illustrations of some of these exotic finds, as well as many manmade artifacts that included feathers, shells, skulls, and many other natural objects, began to be printed by the newly established publishing houses in Western Europe. The illustrated Spanish edition printed in 1521 of Mandeville's '*Travels*', a 14th-century compilation, includes an illustration of a plant which, when ripe, discloses an animal like a sheep, while Conrad Gesner's[3] '*Historia animalium*', published in Zurich between 1551 and 1587, illustrates a many-headed hydra.

Illustrations were, more often than not, based on descriptions rather than observations, and thus in many cases bore little resemblance to the actual appearance of the animal itself. Even the country or place of origin of the artifact or specimen was often unidentified, particularly if items had been traded across vast areas of then, unknown lands and oceans. In modern terms this is similar to some traders in Fiji, who are currently selling items as "Fijian" souvenirs, which actually originate from the Solomon

Islands. The best documented example of this trade is perhaps the golden cowrie, *Cypraea (Lyncina) aurantium,* which was traded across the Pacific to Tahiti from Fiji.[4] The foreword to May and Henry Larson's book on New Caledonia and its peoples states

> *This lustrous golden shell first roused the covetousness of the inhabitants of islands less fortunate than those where it was most often found, Fiji and the neighboring archipelagos. Long, long ago, wars were fought and blood flowed in the paradisiac islands of Oceania, all because of the golden cowry. Then came the first conquerors, the intrepid navigators and the great sailing ships. Greedy for the new, for unknown objects to carry back to relations and friends, to fair admirers, or sometimes also to cabinets of curios—the forerunners of our museums—they were attracted at once by these heavy, highly colored shells which the islanders wore as ornaments, and acquired them in exchange for tobacco, weapons, cloth, or trumpery.*[5]

The tendency to be fascinated by the unfamiliar led to many folk tales, which still surround both the well-known and the unknown. Tales of Bigfoot in North America, the Yeti in the Himalayas and the Loch Ness Monster in Scotland all continue to have their believers, who are convinced that some spark of truth must lie behind these stories. The formation of the International Society of Cryptozoology, founded in Washington D.C. in 1982, encourages scientific understanding by gathering evidence of creatures that may or may not exist.

Public fascination with novelty led to a few taxidermists inventing species, Charles Waterton (1782–1863), being perhaps the most famous but certainly not the only one, was known frequently to abuse the art of taxidermy by producing creatures made up from the parts of others such as the Noctifer, or "Spirit of the Dark Ages", half-bittern and half eagle-owl. His most famous creation was simply entitled "Nondescript" and appears as the frontispiece of Waterton's *Wanderings in South America.*[6] Waterton himself admitted that the creature was a lampoon on the illiberality of the British Treasury, which ruled that he had to pay duty on all the animals imported into Britain by him. The face of the Nondescript represented J. R. Lushington, the Secretary to the Treasury who signed the letter sent to the customs officials.[7]

CABINETS OF CURIOSITIES

The earliest sizable cabinets of curiosities, emerging in the 16th century, (although earlier small collections had existed in the past) were built to house and celebrate strange objects brought back from far-away places with exotic cultures. However, superstitions and the religious dogma that often accompanied collectors resulted in few of these objects retaining their original cultural meanings. Most of the cultural origins, stories and uses of these natural objects have been lost, or reinvented to fit modern cultural trends.

War exacerbated the situation for many objects, which were brought back as trophies as well as casual souvenirs, often without the collector knowing anything about the objects, their meaning or their significance to the communities from which they were removed. However, these same objects may acquire aesthetic, monetary and even new cultural values within the importing communities, blurring the original significance of the objects themselves. Aristotle (384–322 BC) believed that the way to understand the universe was through observation, he also philosophized that all objects have a defining essence, which could best be understood by reason rather than experimentation. These opinions were to condition future philosophers and scientists well into the 18th century.

Most objects were originally collected by individuals: sailors, soldiers or explorers and even missionaries who found that they could supplement their income by selling them on to wealthy collectors. Perhaps the best known of these items was the Precious Wentletrap. Sale of this shell, the most desirable type in Europe in the 18th century, was monopolized by the Dutch, as were most items from the East Indies at the time. The Dutch dealers in Amsterdam are thought to have supplied many of the early cabinets with natural curiosities from the east. This shell was so much in demand that Maria Theresa's husband, Kaiser Franz Stephan, paid 4,000 guilders for one in 1750. The first illustration of the Precious Wentletrap was published in 1705 by Rumphius in '*Amboinsche Rariteitkamer.*'[8]

Such wealthy collectors could enhance their own prestige by showing off this material, or at least allowing access to their collections by their peers, resulting in cabinets of curiosities becoming immensely popular in the royal courts of Europe throughout the 16th and 17th centuries.[9] Typical contents of such cabinets of curiosities included dried fish, some exhibited as proof of mermaids. One placed on exhibition in London in 1775 was found, according to its owner, in the Aegean Sea. Edmund Burke and other eminent scientists had no doubt that it was real.[10] Other examples include horns of mythical creatures such as the unicorn, now known to be the tooth of the male narwhal *Monodon monoceros,*[11] various types and sizes of saws from sawfish, as well as items that, to us, seem more mundane, such as sections from the skin of an elephant, sharks' teeth or oddly shaped pieces of sponge and coral formations. A typical cabinet of curiosities would contain many thousands of objects, depending on the wealth and position of the collector. The following is part of an account by a German traveler who visited the 17th-century collection of John Tradescant (*d.* 1637) in London.

> *In the museum itself we saw a salamander, a chameleon, a pelican, a remora, a lanhado from Africa, a white partridge, a goose which has grown in Scotland on a tree, a flying squirrel, another squirrel like a fish, all kinds of bright coloured birds from India, a number of*

> *things changed into stone, amongst others a piece of human flesh on a bone, gourds, olives, a piece of wood, an ape's head, a cheese, etc., all kinds of shells, the hand of a mermaid, the hand of a mummy, a very natural wax hand under glass, all kinds of precious stones, coins, a picture wrought in feathers, a small piece of wood from the cross of Christ . . .* [12]

These collections enabled collectors to impose an order on the natural world and thus imply power to control the products of creation. The Zymoglyphic Museum website illustrates various Baroque cabinets from across Europe (Figure 2.1).[13]

These collections often contained objects that broke all of the known rules governing the natural world; a species such as the duck-billed platypus *Ornithorhynchus anatinus* is typical. A mammal, which laid eggs, had a duck's bill, a flattened beaver-like tail, webbed feet and venom glands was just too much for Western scientists and it, like many other specimens, was rejected as fake. (It did not help matters, as mentioned earlier, that taxidermists did produce fake animals, designed to mislead the unwary collector).

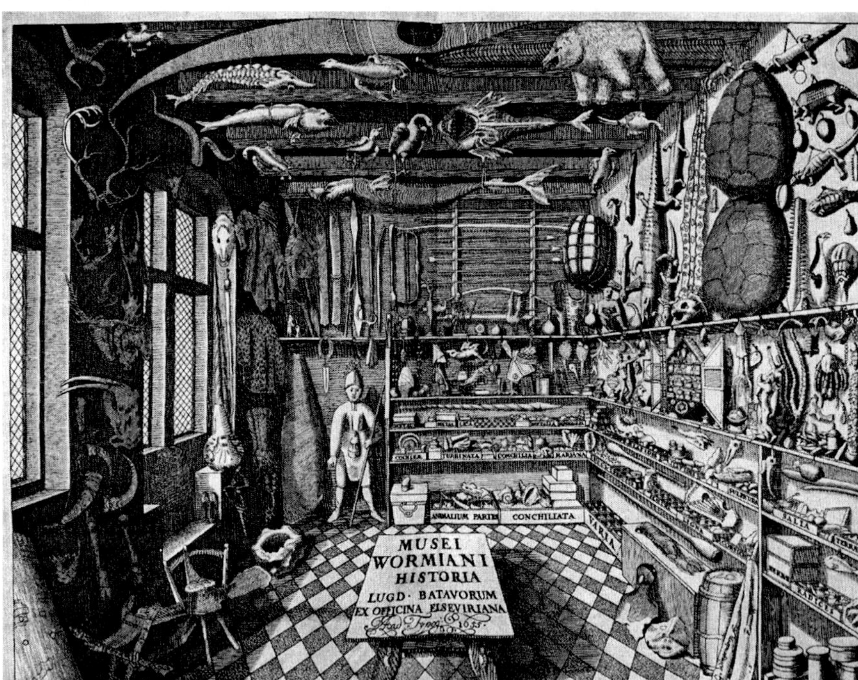

Figure 2.1 Ole Worm's cabinet of curiosities from "Museum Wormianum", 1655. Original source: Smithsonian Libraries.

Far-fetched stories surrounding these animals increased the interest in, and demand for, such specimens.

The science behind the art of taxidermy has advanced dramatically since early attempts to replicate the image of a living animal, which, in many cases, had never been seen other than in a semi-preserved state. Modern images, particularly moving images showing animals in their natural habitat, allow the modern taxidermist to identify those intangible but important factors, which give each individual species its own unique appearance. It is understandable, therefore, that few, if any, of these early preserved objects could be classified by modern terminology as pieces of taxidermy; preserving animals for scientific study and display is very much a modern Western concept. Some of the earliest known extant examples of taxidermy, as most understand it today, date back to the 17th century, for example, the Kungliga Slottet (Royal Palace) and Museums in Stockholm displays a stuffed horse named Streiff, which belonged to Gustav II and was killed in battle in 1632, while the oldest surviving bird known is the Duchess of Richmond's parrot of 1702, which is preserved in Westminster Abbey in London, England. Unfortunately, history does not record who preserved this unique bird. These examples are valued as much for their association with specific people as they are for being examples of very early historic taxidermy.

Early explorers and their crews, like all modern-day tourists, returned with souvenirs of their adventures. From the end of the 15th century to the end of the 18th century, numerous voyagers left Europe to circumnavigate the globe. Many natural objects collected at the time, for example the then rare golden cowrie mentioned previously, often had a far greater cultural significance in its place of origin than in the hands of wealthy collectors. Evidence suggests that the first specimens of the golden cowrie seen in Europe were obtained on Captain Cook's second voyage of circumnavigation in 1772–1775 aboard the *Resolution*.[14]

However, Edward Donovan in 1823[15] reports that the eminent naturalist Sir Joseph Banks observed that this shell was a greatly valued part of the traditional dress of Otaheite [Tahiti]. The pierced shells can still be found in many early collections. Sir Joseph Banks (1743–1820) accompanied Captain Cook on his first circumnavigation on board the bark *Endeavour* 1768–1771. Natural sciences specimens were also collected in the form of objects made from animals and birds such as Maori feather cloaks from New Zealand. A cloak of kiwi *Apteryx* sp. feathers (*kahu kiwi*) now in the British Museum, which probably dates from the 19th century, is an example of museum specimens losing their contextual information, leaving behind only the tangible object. About 40 cloaks were collected on Captain Cook's voyages, but few bear traces of feathers.

Magnificent feather cloaks and headdresses brought back from the Polynesian Islands and South America highlight a wider even and more significant heritage, much of which has been lost, even if subsequently revived by native communities sometimes with differing, if not altogether different, cultural meanings than those originally intended. When the book *'Te Maori: Maori Art from New Zealand Collections'* was published [16] to coincide with the opening of the *'Te Maori'* exhibition in 1984, Maori elders turned their anger on what they considered to be grossly incorrect treatment of the associated *whakapapa* (genealogy) and history.[17]

John Tradescant is considered to have been the first to display publicly in Britain a collection of natural curiosities. A sumptuous catalogue of his collection was published by his son John the Younger (1608–1662) under the title *Musaeum Tradescantianum* in 1656. This collection became the founding collection of the Ashmolean Museum at the University of Oxford, England, after the collection was bequeathed to Elias Ashmole by John the Younger. The whole collection then known as the 'Ark of Lambeth'[18] or 'Tradescant's Ark' was transported to the University of Oxford where erection of a building to house the collection between 1679 and 1683 established the Ashmolean Museum as the first public museum in Britain.

From the 17th to the early 19th centuries, the aristocracy and gentry in Europe continued to collect and display curiosities to their peers. Ralph Thoresby (1658–1724) of Leeds, Yorkshire, England, acquired an extensive collection of natural curiosities, which in its early years was available for inspection only by invitation. However, towards the end of his life, he did allow the more affluent inhabitants of Leeds to visit his private museum. Unfortunately, his museum was dispersed soon after he died and few items from his collection still survive.[19] Increases in shipping and later world trade produced an expanding wealthy merchant class in Britain and elsewhere, followed shortly afterwards by a growing affluent middle class, a product of the Industrial Revolution. This expansion of wealth also extended to the professional classes, with many desiring to enhance their status as modern, free-thinking men and women of the world. Consequently, collections of curiosities became more widespread and popular phenomena among the upper and middle-classes of society. Mary Anning (1799–1847) of Lyme Regis in Dorset, England became "the greatest fossilist the world has ever known".[20] Acknowledged by the Geological Society of London as the first person to discover the *Ichthyosaurus,* a fossil marine reptile. This discovery was made at a time when the bulk of the credit for geological discoveries was received by upper-class London men.

Charles Darwin (1809–1892) was to spark one of the most significant and revolutionary debates among both the scientific and the religious communities with the publication in 1859 of his book *On the Origin of Species by Natural Selection.* Its publication caused uproar in Victorian society and the philosophy behind his observations and ideas still rages as debate in some sectors even today. Ongoing exchange between the scientific community and the Church raised more opportunities for professionals to become involved in establishing natural and scientific collections, without clashing with ingrained religious beliefs.

REGIONAL MUSEUM DEVELOPMENT IN THE 19TH CENTURY: THE EXAMPLE OF YORKSHIRE, ENGLAND

In 19th-century Yorkshire, professionals such as teachers, doctors, scientists and engineers came together with industrialists and businessmen to help educate and expand their own and others' horizons. This coming together took the form of private learned societies in which discussions and demonstrations could take place.[21] Such demonstrations included practical knowledge, such as how to dismantle and clean a rifle, as well as major scientific advancements taking place at the time, for example, demonstrations on the use of electricity. The museum of the Leeds Philosophical and Literary Society, which opened in 1821, was typical of the many regional museums that became established in early 19th-century Britain. The dominant subject, however, was the natural world and the amazing new finds being reported from all corners of the globe. The Leeds Philosophical and Literary Society modeled itself on the Royal Institution in London even instituting a series of Christmas Lectures, similar to those still being given by the Royal Institution to this day.[22]

The subjects of the public lectures reflected the wide interests of members; the first two lectures on record were 'The Effects of Spirituous Liquors on the Human Frame' and 'The Destruction of Herculaneum and Pompeii'. Henry Crowther, curator of the museum, expanded these lectures for schools and in 1901 instituted a Scheme for Scholars Visiting the Museum, soon involving elementary school teachers, and even supplying free transport from the school to the museum. Between 1893 and 1928 Crowther gave over 2,000 lectures to 12,000 teachers and over half a million school children.[23]

These private organizations, most commonly known in Yorkshire as Literary and Philosophical Societies, grew quickly and soon established their own collections and, subsequently, their own permanent museums, reflecting the growing interest in man and his environment. In the early days, acquisition of objects and specimens for public display by learned society museums was simply a matter of receiving haphazard passive donations. However, as collections developed arrangements were made with travelers, seafarers, missionaries and colonial administrators for material to be collected and sent back from abroad in order to increase the offer available for study and for attracting visitors. The considerable collections amassed by these societies in the wake of growing global exploration and colonial expansion were to become the foundation for Yorkshire's major regional museums.[24]

As time went by, these learned societies sponsored a more forward-thinking and systematic approach to their collections, in an effort to display as wide a range of material as funds and opportunity would allow. They acquired more wide-ranging collections of birds and mammals to display the form and diversity of life on Earth. This gave the society museums the opportunity to contrast systematic collections of local species with more exotic examples of similar species. Thus, birds and mammals from Australasia and large numbers of primates from both the Old and New Worlds found their way into these collections. A number of such organizations became established in Yorkshire with societies being founded in Leeds in 1818, Sheffield 1822, Bradford, Whitby and York in 1823, Wakefield and Scarborough in 1827 and Halifax in 1830.[25]

These museums still exist and in most, the collections became the foundation of the region's current publicly funded institutions. Acquisition of these museums and their collections by local government authorities up to 100 years after their foundation, in many cases by default, restructured the necessary funding base for these fast-growing collections, which outpaced the possible revenue sources from local independent societies. The wide-ranging and diverse collections acquired by the literary and philosophical societies left a complex legacy often far beyond the capacity or understanding of the average local authority governing body. In particular, natural history collections often remained underfunded with, in the mid-20th century, important historic items being removed from public display into permanent and often inadequate storage conditions.

SCIENTIFIC COLLECTING EXPEDITIONS

In the rest of Europe, the power of collections-based knowledge also extended into the realm of geo-politics and policy. Collections were even sometimes used as instruments of the state. For example, in the 19th century, the Russian imperial authorities created a collection of exotic taxidermy in St Petersburg that included a kangaroo and an anaconda. It was assembled exclusively for the purpose of enlightening and enhancing the status of Russian sailors and other official travelers representing Russian interests. They were to appear cultured and knowledgeable about the known world, its peoples, cultures and native faunas, rather than to express amazement at the sight of exotic environments, which might indicate a lack of education and sophistication.

In the final quarter of the 19th century and the early years of the 20th century, countries varied in their economic capacity, political stability and influence, in terms of their ability to mount expensive scientific expeditions to gather material from the field for their museum collections.

In Britain, most expeditions were only undertaken under the auspices of the larger universities, but in other countries such as Germany and the Austro-Hungarian Empire, imperial support helped to fund collecting trips specifically for museums. For example, the *Naturhistorisches Museum* in Vienna contains material originating from the Imperial Hapsburg collection and from 19th-century voyages of discovery, such as that of the *Novara* world circumnavigation, which attracted imperial patronage.

The devastating effects of two world wars in the 20th century destroyed some of the material collected earlier and restricted funding and opportunities for collecting by most European museums. However, many American institutions and wealthy individuals had the resources to continue collecting throughout the 20th century.

Both world wars helped dismantle the empires of Great Britain and other European states. However, in contrast to the damage usually caused by colonization, foundation of museums by colonizing and occupying powers often left a positive legacy of institutions that have subsequently flourished post-independence and have helped to build new identities.

VANISHING HERDS

In Britain, insufficient resources and a lack of specialist staff, coupled with frequently cursory documentation, some lost due to war damage, led to many specimens, especially larger ones, being put at risk by lack of research into their possible importance. Intangible heritage connected with museum objects adds richness to them. Others, that have stories, which have not been fully documented, provide an untapped opportunity to increase their heritage value. These stories may lead to important discoveries about the animal itself, the collector, conservator or taxidermist, or the area of the originating culture and its beliefs.

For example, the Leeds City Museum accession register for 1862 and 1863 contains the following entry "Capt. Edmund Smyth, Bengal; very fine specimen of the Yak *Bos grunniens*". This limited information was all that was deemed necessary in the 1860s; the card index relating to this specimen stated that it was an Indian Yak. Full details of the acquisition of this animal turned up rather unexpectedly over 120 years later when the author Charles Allen, who was researching a book on early explorers and pioneers in the Himalayas, came across a diary entry of a hunting trip into the Garhwal Himalayas. Edmund Smyth (1823–1911), immortalized as Crab Jones in the novel *Tom Brown's Schooldays,*[26] was the first European to go mountaineering in the Himalayas. In 1862 he led a hunting party to the sources of the Tsangpo-Brahmaputra. The yak itself was shot close to the shores of the Sacred Lake of Manasarover in Tibet. Both Edmund Smyth and his friend and fellow hunter the Honorable Robert Drummond were government officers, and Robert Drummond was the younger son of the Eighth Viscount Strathallan. The hunting trip was in defiance of a government ban on entering Tibet, and it was particularly important to Robert Drummond that these illicit hunting trips were kept secret. This resulted in the rather vague provenance when the yak was donated to the collections at Leeds.[27]

Similarly, the "Leeds Man-eating Tiger" is far more important in relation to Major-General Sir Charles Reid, G. C. B. of H. M. 2nd Goorkahs (Gurkhas), (Sirmoor Rifles) who shot the animal. Major-General Sir Charles Reid distinguished himself and his regiment at the siege of Delhi in 1857, by holding fast in their redoubt and defending the base against much larger forces. Paradoxically, there was little or no evidence of the tiger having actually killed anyone and the "man-eating" description seems to reflect an early example of attention-grabbing sensationalism by museum staff.

By the 1980s in Britain, the idea of collecting, and thus killing, any animal or plant for any reason became highly unpopular. This was mainly due to pressure from conservation organizations, including some wildlife trusts that believed at the time that all animals and plants could be identified from photographs. Leaflets began to appear asking those who visited the countryside to "only take photographs and only leave footprints". This resulted in museums and the scientific community being placed under pressure by conservationists, the general public and political leaders. This pressure had already resulted in the Protection of Birds Act (1954) and the Wildlife and Countryside Act (1981, amended 1985 and 1991). These new laws placed many museums and their collections under the public spotlight, with conservationists rightly highlighting some of the poor displays and working practices in some museums.[28] In contrast, however, to changing attitudes in Britain over recent decades, in much of continental Europe and other parts of the world, collecting and shooting for sport (sometimes of rare species and in large quantities) still continues as an acceptable activity. Debate rages on this topic, as well as between conservationists and communities, which are striving to protect their traditional ways of life, even though these may involve the hunting of otherwise protected species.

In 1988, the author published a short paper entitled *Vanishing Herds—large mammals in museum collections?* [20] This paper highlighted what at the time was the plight of large mammals in museums in the British Isles. From the 1970s large mammals and non-British birds held by a number of museums and other institutions in Britain were being discarded as having little or no perceived value, monetary or otherwise. Collections were rationalized to reflect financial and political position of a nation that, in the day, gave little thought to culture in any form. Thus, natural science collections suffered the same fate that many ethnographic collections had suffered more than a decade earlier, when numerous museums disposed of their holdings. In some extreme cases collections were given, sold or exchanged to unscrupulous dealers who removed much of the unwanted material from museum collections, only to sell items to museums overseas for significant sums. Some of these pieces would now be worth many thousands of dollars if placed on the open market.

Lack of concern for the cultural underpinnings of objects, particularly non-local and non-British material, also resulted in the destruction of many displays and dioramas within Britain. Some European institutions followed Britain's lead, resulting in a reduction of the cultural value of natural science collections, in particular, their holdings of historic taxidermy. In a chapter by Eirik Granqvist, *Case study 4: a museum is dismantling its habitat dioramas*,[29] he laments the destruction of dioramas in the name of modernization. The last words of this well-illustrated chapter states "Stuffed animals are not just old stuffed animals, they have an important story to tell".

THE COST OF COLLECTING

In Britain, changing political fortunes and the advent of new governments with new priorities and policies that stressed 'value for money' affected all public services, including publicly funded museums. The government even commissioned a controversial report quantifying the costs involved with collecting and storage, in order to frame this work in purely financial terms.[30]

Many smaller and even some larger institutions, often only employed non-specialist staff to manage and display their natural science material, sometimes with little or no input from knowledgeable volunteers or professionals in other better-staffed institutions. In Leeds for example, in the absence of a specialist geologist in the mid-1960s, the geological collections became the responsibility of the museum's archaeologist/numismatist. In British museums, unlike those in many other parts of the world, humans are not included as part of the natural sciences. Anthropological history, biology, tools and objects of worship are all considered as separate and thus man may be covered by archaeologists, anthropologists, ethnographers or social historians, depending on the main subject areas covered, the interests of senior management and the employing authorities. This lack of specialist knowledge about disciplines such as the natural sciences often resulted in museums being asked to revisit and rewrite their collecting policies. Museums were asked why they collected, what value the collections had, both scientific and financial, and to justify their holdings. Emphasis was placed on the need to rationalize and dispose of any unused and unwanted material.

This pressure led to research being undertaken by Manchester Museum on the "Value and Valuation of Natural Science Collections" with a resulting conference and publication. Within this, David T. D. Clarke in his chapter on museum disposals[31] talked about an anonymous county museum service, which decided to abandon its natural history, unfortunately unfashionable outside safari parks. All of the material was subsequently transferred to the local natural history society, thus placing the material at potential risk in a body without museum safeguards and standards.[32]

In a recent article by Adam Schoon,[33] a member of staff of Tennants Auctioneers of Leyburn, England, states that in the 1960s John Willett, a founding member of the British Deer Society, acquired at no cost the entire taxidermy collection of Norwich Castle Museum, which simply wanted to

Table 2.1 1822–1823 registry entries from Leeds Museum

10 British Birds	Mr. Robert Leyland
The rarer native plants about Halifax	Corresponding member
80 mosses from Halifax	Halifax
A collection of foreign Ericae	
A badger	
and many single articles of value	

get rid of it. The extent to which larger institutions can be expected to take up unwanted, but important material from smaller museums is a matter of debate. Museums themselves have an ongoing dialog around problems associated with long-term storage and display of 19th-century taxidermy dioramas, as well as the costs of housing the often extensive holdings of other smaller natural science institutions.

Early accession registers contain many hundreds, if not thousands, of items, which at first glance have little or no recognizable information attached. For example, the first entry in the registers of the Leeds Philosophical and Literary Society Museum[34] are seen in Table 2.1.

The last item on the list is perhaps the most elusive, and best aligned with the idea of intangible natural heritage. What was meant by this statement? Of value to whom? Perhaps scientists, collectors or native peoples? Or was this simply a reflection of their monetary value? All who work in museums have acknowledged problems associated with such collections, and arguments around a long-term viable future for scientifically important museum material needed to be resolved.

Previously, collection registers presumed that the curator's knowledge would be passed down through generations thus the abbreviations and shortcuts used by museum staff, many of whom were volunteers, were considered adequate at the time. Additionally the constant exchanges of material among institutions such as the Natural History Museum in London, Leeds and the Liverpool Free Museum resulted in large quantities of material becoming detached from tangible information sources such as labels or collectors' letters. Thus, a substantial amount of material in some museum collections, often collected by amateur naturalists using their personal coding systems, could result in the total loss of the scientific value contained within this material.

Considerable research may be needed to reunite tangible objects with their intangible information. For example, in the Leeds collections a local dipterist used the code letter A to indicate his home site Adel. A problem arose when he moved to Austwick in the Yorkshire Dales and continued to use the code A for his new location. He knew the date on which he moved, but it took many hours of research to establish the date range for each of the two sites.

STORAGE AND DOCUMENTATION

An inherent tension exists between the fact that most natural history collections and documentation are housed in Western institutions, while some of the most pressing developmental and environmental needs are in developing countries. Demands from developing nations for access to information relating to their own flora and fauna held within Western collections, highlighted limitations in documentation of the extensive holdings in larger museums, as well as a dearth of specialist staff available. If access to these major sources of environmental data was to be available to outside researchers with the required knowledge, major improvements needed to be made to the environmental and physical storage of these collections and new, more sophisticated documentation systems needed to be established. Within Britain, this debate started in 1973 as a result of the Wright Report on non-national museums.

> *9.7 In order to be able to interpret and communicate knowledge effectively, a museum must first have detailed and accurate information about the objects in its collection. Museums can provide an efficient service only if their information resources are readily available and if their records are revised as a continuing process. Accurate and well documented information also has a vital part to play as a means of collection control: in addition to making it easier to determine collecting patterns and supervise conservation treatment, it simplifies the mundane task of stock-taking and audit.*[35]

In 1985 in Britain, the Museum Documentation Association published the first in a series of papers on planning and implementing the documentation of museum collections,[36] followed in 1992 by the then Museums and Galleries Commission, which published standards for the Care of Biological Collections in Museums[37] and, in 1993, those for the Care of Geological Collections.[38] Museums hold vast quantities of valuable data, often within or attached to the tangible component, the specimens themselves. Even with modern scientific equipment, computers and the World Wide Web, the museum sector still struggles with making this information widely available. Manuscripts, illustrations, diaries, letters and other unpublished documents, as well as scientific reports may reflect intangible heritage and present an opportunity to know more about each specimen. Where they are available, such additional sources would help to expand our knowledge, provided that the connection between the object and the archived information can be reestablished. Within Britain, material such as letters, diaries, family documents and records, estate records and similar ephemera are usually housed within the County Archives Service, which is completely separate from most of our museum services. Thus, important tangible material and intangible information relating to specimens held within our museums can be held in different institutions and separate geographical areas. This often results in the staff looking after specific collections being unaware even of the existence of the sources of important documentation relating to specific objects.

Over about the past 20 years, most museums and equivalent archive departments in Britain have made attempts to digitize the data held in their collections, trying to combine tangible and intangible information into a single useable resource. Most museums, however, have no long-term guaranteed funding to ensure that these data can be easily accessed online by all interested parties, wherever they reside.

Many parts of the world have online data available for searching and browsing. For example, Australia has Collections Australia Network (CAN),[39] New Zealand has eHive[40] and The Smithsonian Institution in the U.S. has its own Central Repository.[41] Within Britain, nationally we have FENSCORE, The Federation for Natural Sciences Collections Research, an online searchable database of collections held within museums and similar institutions.[42] However, little nationally based work has been undertaken on listing individual items within collections online.

Some attempts to resolve these issues have been made in recent years in Britain. First, in 1988, came a Registration Scheme for all *bona fide* museums. In order to achieve Registration, museums had to comply with minimum standards relating to their constitution and staffing, as well as the documentation of, and access to, their collections. This was followed in 1997 by a scheme of Designation of Nationally and Internationally Important collections, initiated by what was then the Museums and Galleries Commission, now run by the Museums, Libraries and Archives (MLA) Council.[43] Designation came with some short-term funding packages. More recently, as a result of a task force report in 2000,[44] to establish a sounder financial situation for regionally significant museums, the government established and developed a series of Regional Hubs. These Hubs are based around one or more major museums within each region. They receive extra funding from government through the Renaissance in the Regions program principally to develop the educational role of all museums within each region. Registration, the minimum standards scheme established by the United Kingdom (UK) authorities was, with local variations, subsequently adopted by several other European countries. Standards schemes have also been developed in other parts of the world, such as the Museum Accreditation Scheme run by the American Association of Museums.

Internet communication is, by its very nature, transitory. Archiving electronic documents occurs erratically, if at all, which may result in their availability to researchers growing more limited over time. However, the low costs involved, compared to traditional publishing and distribution channels, have resulted in a propensity by many to publish exclusively online. News items so important for those wishing to be kept up to date are often available for reference for only a short-time period. Online journals, privately published papers, blogs and many other formats all jostle for the top slots allocated by individual search engines. Papers published only in a Web format can regularly change or disappear altogether, lessening their ability to make a long-term contribution to the state of knowledge of a subject. However, by contrast, the long- or well-established scientific journals with established archives of past publications and the many archives, which republish papers for downloading, can be a boon to researchers.

RECENT AUCTION SALES OF HISTORIC TAXIDERMY

Valuing the scientific and cultural importance of objects is both a tangible and intangible problem facing all museums. Collecting museum specimens is often a costly exercise; most institutions benefit from the time and expertise of many hundreds of collectors who have, over the years, contributed objects at their own expense. Museums who accept these items into their care are undertaking a long-term commitment to conserve and store them, as well as making a tacit agreement to render them accessible for future research. The presence of a specific object within a museum's collection may, or may not be, justified on purely aesthetic grounds. Mounted trophy heads are a prime example. They are often large, difficult to display and take up a great deal of valuable storage space. However, even these unloved items can have both scientific and monetary value.

In 1997, Manchester Museum made an attempt to estimate the cost of collecting[45] natural science material, also highlighting the monetary value that could be placed on museum holdings. But even after this publication, which supported the earlier Cost of Collecting report,[46] many museums in Britain and Europe still placed little value on their natural science holdings, from either a monetary or scientific perspective.

Contrary to earlier trends, in recent years the monetary value of specific individual pieces of taxidermy has soared. In Britain in the 1950s and 1960s, old cases containing taxidermy had little monetary value and could be acquired easily at very little cost. In the 1980s and 1990s, few auction rooms would sell this type of material for fear of falling foul of the complex conservation regulations or hostile public opinion. The recent resurgence of interest is testified by the results of a number of sales by auction of private trophy collections. For example, Tennants Auctioneers of Leyburn, Yorkshire, England, have sold a number of such private trophy collections in recent years. In 2005 they sold the collection of Kenneth Whitehead for a combined total of £470,000. In July 2006 the collection of trophies amassed by Christian Oswald, removed from a wooden hunting chalet in Bavaria, was sold. This collection included the antlers of the extinct Giant Elk *Megaloceras giganteus* with an estimated sale price of £20–30,000. The importation of much of this material from continental European sources must reflect the higher prices paid in Britain at auction for individual specimens in such collections.

The increase in prices for taxidermied specimens is an identifiable trend. In the autumn of 2007, Tennants sold the collections of Ken Aldridge and

Richard Grassmann (1930–2006). In July 2008 the collection of Jacques and Micheline Henrijean of Belgium was sold. This large collection included a magnificent specimen of the Polar Bear *Ursus maritimus,* which sold for the hammer price of £11,500 (with a buyer's commission of 15%, the total paid for this single item was £13,225). The antlers of the extinct Schomburgk's deer *Cervus schomburgki* reached a hammer price of £5,500, an Alaskan brown bear *Ursus arctos middendorfi* reached £8,800, and two mounted jaguars *Panthera onca* brought £7,500 and £4,200, respectively. Another polar bear was sold for £6,200. In the same sale was a small Giant

Table 2.2 Hammer Price of Rhino Heads Sold at Tennants 2008–2010

No.	Species	Material	Collection information	Price
		April 2008		
804	*Diceros bicornis*	Mounted head	ca. 1900	£29,000
		November 2008		
1235	*Diceros bicornis*	Both horns	Northern Guaso, Nyiro Rowland Ward	£19,000
1255	*Diceros bicornis*	Mounted head	Gerrards, London	£15,000
1282	*Diceros bicornis*	Both horns	H. Burton, London	£20,000
1310	*Ceratotherium sinum*	Mounted head	Sudan 1921 Rowland Ward	£65,000
1343	*Diceros bicornis*	Mounted head	Rowland Ward	£7,000
1413	*Diceros bicornis*	Both horns	British Eeast Africa Rowland Ward	£16,000
		Summer Sale July 2010		
1139	*Diceros bicornis*	Anterior horn	ca. 1910	£19,000
1146	*Diceros bicornis*	Both horns	Namibia pre-1947	£1,900
1153	*Diceros bicornis*	Both horns	Shot 08/24/.1907 by W. A. Bowring	£42,000
1163	*Diceros bicornis*	Both horns	Limpopo region ca. 1930 (Rowland Ward)	£70,000
1176	*Diceros bicornis*	Both horns	ca. 1913	£41,000
1195	*Diceros bicornis*	Both horns	German-East Africa. Shot ca. 1933 Max-Dietrich Gaudchau	£46,000
1162	*Ceratotherium sinum*	Both horns	Limpopo region ca. 1930 (Rowland Ward)	£106,000
1211	*Ceratotherium sinum*	Anterior horn	WSB 1907	£43,000
1221	*Ceratotherium sinum*	Mounted head	South Africa	Not Sold

Clam 59 × 62 cm, which brought a hammer price of £750, and a Galapagos Giant Tortoise *Geochelone elephantopus* collected in circa 1909, which came in at £4,000.

Rhinoceros horn, in particular, has attracted high prices at auction, as demonstrated by results from the sales catalogues published following each sale date.

Traditional Chinese medicine places great value on rhino horn as a cure for many ailments, including headaches, hallucinations, typhoid, fever, rheumatism and gout (Figure 2.2). Some other cultures still consider items made from rhino horn to be prestige items. For instance, daggers called "Jambiya" with rhino horn handles are presented to Yemeni boys at age 12 as a sign of manhood. Thus, objects made from such exotic materials can bring with them added and much-valued status.

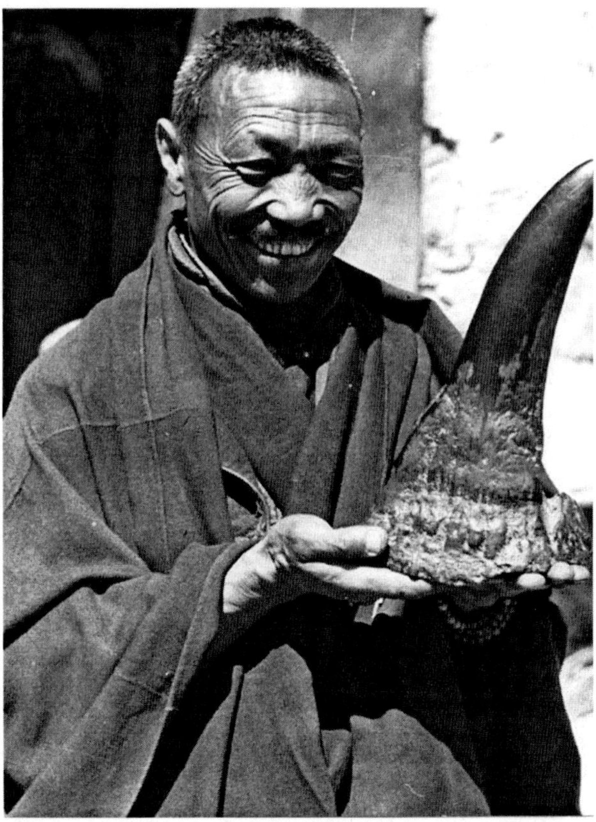

Figure 2.2 Rhinoceros horn has been important to traditional medicine in Asia for a considerable time. "*Umgebung von Samyeh, Nashorn*" (Tibetan Monk with Rhinoceros Horn), 1938, by Ernst Schäfer, courtesy of *Deutsches Bundesarchiv* (German Federal Archive), *Bild* 135-S-15–17–39.

These recent high auction prices may also reflect the growth of museums in richer developing countries, each trying to acquire extinct and endangered material for their own collections, with new fresh material unavailable due to the international conservation laws, conventions and trading agreements such as CITES.[47] It is possible that these sales may also represent, to some people, a legal way of circumventing the rules restricting the international trade in endangered species. Horns sold at auction without supporting information can, in fact, bring as high a price, or higher, than fully mounted heads. However, there are still a number of private collectors of taxidermy and trophies some of whom acquire pieces for decoration and display in private or commercial premises. Which of the above are the cause of this relatively recent high demand for such items is still open to debate. Unfortunately, auction houses in Britain rarely reveal the name of the purchaser, and so it is usually impossible to know who is purchasing or why such items are being purchased. Rules may vary between countries, but the London auction houses have bases in both Europe and America and in-house conditions of sale are usually similar.

Even material that is not sold off can languish in long-term storage in perpetuity. Most museums and institutions in the world hold at least some valuable items, which may be neglected, even if stored in good conditions, and are rarely if ever seen by the general public. A good example of the plight of such collections is that of the Echalax Bird Collection.[48] The collection of Lt. Col. C. T. Echalax was donated to the people of Liverpool in 1907. The collection became a central part of the displays at the then newly established Waterloo Museum, which opened in 1908. Bombing, vandalism, several moves and the vagaries of time have taken their toll on the collection and, unfortunately, it is now held in a basement store in Crosby, in the care of the local authority. The tendency for institutions, and curators themselves, to place unfashionable or troublesome material into long-term storage is regrettable, but time, lack of interest by both staff and outside parties and the costs involved can all result in this pragmatic option.

Institutions holding paintings, or even single decorative art objects such as jewelry, glass or ceramics, valued in excess of £10,000 would almost certainly ensure that such material was visible by the public, or at least held in safe secure storage, simply because of its perceived monetary value, as distinct from any scientific or artistic value it might possess. Establishing the cultural value of an object can result in a dramatic change in the fortunes of such individual objects.

For example, the intense media interest in the skeleton of the famous racehorse Phar Lap, currently held in the Museum of New Zealand/Te Papa Tongarewa (Te Papa), resulted from a request to exhibit the skeleton in Victoria, Australia for the 150th anniversary of the Melbourne Cup. This animal is held in high cultural esteem by the peoples of two different nations, with various parts of the animal now resting in several different museums. Its skeleton is in Te Papa in Wellington, the mounted hide is displayed in Melbourne Museum and its heart is in the National Museum of Australia in Canberra. The remains of few other animals could engender such passion.[49]

THE ART OF TAXIDERMY

In recent years, a great deal of new interest has also developed around the history of taxidermy and taxidermists. The recently formed British Historical Taxidermy Society was established by a small group of private collectors to try to save at least some of the most prized examples of the art of taxidermy from being sold overseas. Additionally, in recent years a number of books have been published by enthusiastic specialists on individual taxidermists and their work. Books on Rowland Ward,[50] Edward Gerrard and Sons,[51] including a Digital Taxidermy Archive, George W. Quatremain,[52] Van Ingen and Van Ingen[53] and Walter Potter and his Museum of Curious Taxidermy, have all proved to be popular. In Helen Fothergill's review of the book on Walter Potter, she states "it is a wonderful reminder of what we have lost as an 'entity' and perhaps a warning to the museum world as a whole that we are in danger of not seeing the wood for the trees". [54] Fothergill also states that "these wonderful eclectic groups—don't sit comfortably within our collecting policies".[55] The Potter Museum was sold at auction in 2003, and the collection dispersed to private collectors.

Due to the work of Damien Hirst and other contemporary artists, the art world has also recently taken an interest in taxidermy. Some artists have even been trained in the art of taxidermy by staff at the National Museums of Scotland. This expanding interest by the art world in the subject does little, however, for taxidermists in general. In Britain, very few museums still employ taxidermists on their staff, preferring to follow governmental guidelines to outsource this expertise on an *ad hoc* basis. This has resulted in a fall in membership of the Guild of Taxidermists, the UK professional body, by a third in recent years. The membership profile also indicates an older generation of independent professional taxidermists who are now mainly employed mounting hunting trophies and family pets. The future looks bleak in the UK, as at present there are no training courses.[56] Although recent visitor studies suggest that the public, in particular children, enjoy visiting museums with taxidermy on display, there is still a widespread perception amongst museum managers and governing bodies that the British public take a particular dislike to taxidermy. These contradictions are hard to explain, and it would be interesting to explore this issue more fully. An international study on the public perceptions of taxidermy in museums may clarify the situation and help to enhance the status of taxidermy within museums. This may also enable museum curators to make better informed decisions on the future conservation and status of historic taxidermy in particular.

Installations by contemporary artists can be a new and interesting way of displaying both modern and historic taxidermy. However, careful assessment may be required to ensure that installations ensure a positive public reaction to the material being displayed. A number of well-known modern artists have worked with and around traditional displays of taxidermy with great success. For example, Kate Hammersley's exhibition "*Order & Chaos*" reinterprets notions of classification and order, blurring distinctions between the traditional and contemporary in both the arts and sciences. Working in drawing and sculpture her practice is to reinterpret perceptions of truth in relation to natural history museum objects.[57]

Preservation techniques used by Damien Hirst, with whole sharks and sections of calves preserved in formaldehyde, owes more to the many back-room technicians who care for spirit collections used for taxonomic and scientific study, than to professional taxidermists. However, his success as an internationally important artist cannot be dismissed.[58]

THE INTERNATIONAL DIMENSION

Eirik Granqvist[59] lamented the destruction and possible fate of individual specimens removed from dioramas and pointed out the difficulty of drawing a line between those items, which are purely of natural history interest and those of cultural importance. As covered earlier in this chapter, some individual specimens can have a cultural importance well beyond the confines of individual museums or even individual nations. Museums throughout the world contain many millions of specimens, most of which originated outside the national boundaries of the countries in which they are housed. The absence of cultural information being attached to specific specimens does not mean that they have no cultural importance in their country of origin. Every curator, no matter how knowledgeable, has only limited experience and cannot know for certain, or be able to establish beyond doubt, whether or not objects placed in their care have any spiritual or cultural importance outside that of their own national understanding.

Formation of the Working Group on the Art of Taxidermy and Its Cultural Heritage Importance by the International Council of Museums International Committee for Museums and Collections of Natural History (ICOM NATHIST) at its meeting in Jakobstad, Finland, in 2005 was a way of highlighting the international importance of taxidermy not just in Europe but throughout the world. The members of the Working Group believe that material held in museums belongs to the wider community and that cultural aspects should also be taken into account when changing the appearance of large pieces of taxidermy, displaying, interpreting or even deaccessioning or destroying such items.

THE FUTURE

International Loans and Exchanges

Development of research into the intangible aspects of natural museum objects, and display of these items outside their home institutions, have both resulted in an increase in natural science material crossing international borders. Customs restrictions, Convention on International Trade in Endangered Species of Wild Fauna and Flora (CITES) regulations, biological control and even the legal ownership of museum items can sometimes make it difficult for institutions to share the tangible object and, by extension, its associated intangible stories. Data associated with specimens should, however, be freely available within the scientific community, irrespective of location, so an element of trust is warranted. Ownership disputes are a particular problem, causing museums to worry that loans of some material to overseas institutions may be seized, either by courts in the destination country or by authorities within any country through which they may transit. The international community needs to find and establish robust guidelines for the safe loan and guaranteed return of materials between museums across international borders. Several countries, including the UK, have recently adopted anti-seizure legislation, and others are planning to do so in order to facilitate major loans between museums across international borders.

DNA and Intellectual Property Rights (IPR)

Recent advances in DNA research, cloning and the production of new artificial life forms may eventually affect museums. Museums have an opportunity to become more proactive around the many scientific issues associated with these developments, especially when relating their collections to the tangible and intangible aspects of science and cultural heritage. Best practice in international cooperation is exemplified by the scientific community's establishment of internationally funded databanks such as the Biobanking and Biorepositories set up for seed conservation, and counter to an unfortunate trend by some institutions, in particular those copyrighting data, to consider biological information as personal or institutional assets.

International cooperation would benefit other fields such as DNA research. Museums throughout the world hold examples of extinct animals and plants, and this material may be the only source of important biological data. This is more true now that science is pushing new boundaries with cloning. The cloning of Dolly the sheep in 1996 pushed the boundaries of science into a new era of scientific endeavor. Attempts have already been made to clone DNA from museum specimens of the extinct Tasmanian Tiger *Thylacinus cynocephalus*, but the material proved to be too badly degraded. The DNA of other animals for which frozen tissue exists such as the Pyrenean Ibex *Capra pyrenaica* or even the mammoth may prove to be less fragmentary. Research undertaken by Prof Michi Hofreiter, a member of York University's BioArCh research group, has extracted DNA from a mammoth and reconstructed the animal's hemoglobin. This work demonstrated that mammoth hemoglobin released oxygen at low temperatures much more efficiently than modern elephants, allowing them to survive in colder climatic conditions.[60] Research is also taking place in the U.S. to develop 'synthetic life.' The team, led by J. Craig Venter, has developed the first self-replicating bacteria cell controlled by a synthetic genome.[61]

Museums are generally scientifically and technically ill-equipped to take advantage of such advances. Rarely do collections of plants, for example, include information on any medicinal, agricultural or domestic uses they may have had in their country of origin. Such collections are usually only valued for their taxonomic status rather than their scientific or cultural potential.

Climate Change

Museum collections also have a role to play helping track introduction, dispersal and impact of new alien species. Easy and widespread travel and trade have left many regions open to the introduction of pests, with the potential of devastating effects on health, agriculture and native species. Erratic global weather patterns have encouraged the spread of species across long-established natural ecological barriers. For example, warmer seas have resulted in the coastal waters of the Hawaiian Islands being clogged by seaweeds covering beaches and coastal canals and Japanese seaweed is changing the ecology of the sea coasts of Nova Scotia in Canada. An American jellyfish now inhabits the Black Sea. Brazilian earthworms have reached the Philippines. Over 600 species of non-native plants and animals, including bullfrogs and Mediterranean grasses, now inhabit the Sonoran Desert of Arizona and Mexico.[62]

Although some invasive species that have been brought to the attention of specialists are being monitored, many pests are highly likely to cross ecological barriers, unnoticed until they become established and difficult to eradicate. Such ecological disasters already plague many smaller island groups, such as the Pacific Islands as well as larger land masses like Australia and New Zealand, which have had to contend with major infestations due to long histories of alien introductions.

Along with other members of the scientific community, museums can help monitor alien adventives, as staff are uniquely placed to access historic and recent collections. That mystery animal or plant brought in by a member of the public for identification may be a very important new, and possibly dangerous, alien introduction. Many species have been identified as new to a specific region or even, in a few cases, new to science, when a specimen was found hidden among unidentified or misidentified material. These finds may be evidence of unnoticed alien introductions, providing information on the length of time the species survived, possibly even a record of the damage they did. These patterns are able to increase our understanding of the conditions under which species enter and become established in new ecological systems.

The Monetary Value of Museum Collections

Funding for museums in general, and collections research in particular, has rarely been a priority for local and national governments. Resourcing museums must continually compete with healthcare, education and other necessities. The 2008 global financial crisis and its subsequent effects almost certainly hit many museums and other institutions hard, if only in the medium term. In many countries, both public and private sector funding remains in short supply and funding sources may, in the future, become yet more difficult to find. Strategies may need to be adopted to ensure the long-term survival of museums and their collections. Valuing material scientifically and monetarily is becoming a priority.

It may be controversial to suggest that museums should know the monetary value of the collections in their care, as it could be argued that such material is irreplaceable and therefore beyond value. However, all information has a monetary worth. Even with the best storage conditions and practice in place, institutions may still be required to justify their holdings of specific objects or collections. Many museums overlook the possibility that their collections may have some commercial value, even if it turns out to be more esoteric then actual. If natural science collections are to develop, then museums must also establish their importance as resources for data and heritage.

Care of intangible natural heritage requires a fundamental reorganization of the conceptual business of museums; it cannot simply be an addendum to current practice. The work of ICOM over recent years has highlighted the importance of both the tangible and intangible in cultural terms. The ICOM initiatives have also highlighted the importance of the intangible in the development of understanding between peoples with different cultural backgrounds. If this work is to be carried forward into the future, then ICOM needs to have strategies in place to enable it, if so required, to place appropriate pressures on governments to ensure that culture in all its aspects is placed higher on governmental agendas. Staff in natural science departments as well as in other disciplines within museums may be required to improve their own advocacy skills, if they are to compete for limited finances with the more high-profile arts and humanities subjects. In the 19th century, the natural sciences were as highly esteemed as the fine arts. Compare, for example, the identical, palatial buildings of the natural history museum and art gallery in Vienna. In many Western countries, however, arts and humanities collections have continued to acquire increasing prestige throughout the 20th century, at the expense of natural science collections. In the opening years of the 21st century, it is contemporary art that has become the unquestionably fashionable museum discipline. Such vagaries of taste are highly dangerous for museum collections, which need to survive and develop in all subject areas, or future generations may be left with no tangible heritage and only the intangible roots of a previously highly respected and influential part of a once-thriving element of everyone's national culture.

NOTES

1. Robert Huxley, "Challenging the Dogma: Classifying and Describing the Natural World", in *Enlightenment. Discovering the World in the Eighteenth Century*, ed. Kim Sloan (London: The British Museum, 2003), 70–79.
2. John Hale, "'A World Elsewhere': Geographical Horizons and Mental Horizons", in *The Age of Renaissance,* ed. Denys Hay (London: Thames and Hudson, 1967), 317–345.
3. Conrad Gesner, *Historia animalium, II Quadrupedibus oviparus* (Zurich: Froschauer, 1554), accessed February 23, 2011, http://archive.nlm.nih.gov/proj/ttp/flash/gesner/gesner.html.
4. Peter Dance, *Rare Shells* (London: Faber and Faber, 1969).
5. May Larson and Henry Larson, *The Golden Cowrie: New Caledonia: Its People and Places* (Edinburgh: Oliver and Boyd, 1961).
6. Charles Waterton, *Wanderings in South America* (London, 1828), frontispiece.
7. Peter Dance, *Animal Fakes and Frauds* (Maidenhead: Sampson Low, 1975).
8. Peter Dance, *A History of Shell Collecting* (Leiden: E. J. Brill/Dr W. Backhuys, 1986).
9. Adrian Norris, "Presidential address—collectors and collections", *Naturalist* 110 (1985): 125–132.
10. Gwen Benwell and Arthur Waugh, *Sea Enchantress: The Tale of the Mermaid and Her Kin* (London: Hutchinson Press, 1961).
11. The Danish zoologist Ole Worm (1588–1655) recorded in 1638 that a very valuable trade was established importing narwhal tusks from the north and selling them as the horn of the legendary unicorn.
12. See Arthur MacGregor (ed.), *Tradescant's Rarities* (Oxford: Ashmolean Museum, 2010, digital reprint of the 1983 publication).
13. "The Zygomorphic Museum", accessed February 23, 2010, http://www.zymoglyphic.org/exhibits/baroquemuseums.html.
14. Peter Dance, *Rare Shells.*
15. Edward Donovan, *Naturalists' Repository*, vol. 1 (London, Simpkin & Marshall, 1823).
16. Sidney Moko Mead (ed.), *Te Maori: Maori Art from New Zealand Collections* (New York: Harry N. Adams, 1984).
17. Flora E. S. Kaplan, *Museums and the Making of "Ourselves"–The Role of Objects in National Identity* (Leicester: Leicester University Press, 1994).
18. Alma S. Wittlin, *The Museum: Its History and Its Tasks in Education* (London: Routledge and Kegan Paul, 1949).
19. Ralph Thoresby, *Museum Thoresbyanum, or A Catalogue of His Museum, with the Curiosities Natural and Artificial, and the Antiquities; Particularly the Roman, British, Saxon, Danish, Norman and Scotch Coins, with Modern Medals* (London: 1715).

20. Don Brown, *Rare Treasure: Mary Anning and Her Remarkable Discoveries* (New York: Houghton Mifflin Co., 2003).
21. Peter Brears, *Of Curiosities and Rare Things: The Story of Leeds City Museums* (Leeds: The Friends of Leeds City Museum, 1989).
22. Clarke, E. Kitson, *Leeds Philosophical and Literary Society: History of One Hundred Years* (Leeds: Leeds Philosophical and Literary Society, 1924).
23. Adrian Norris and Adam White, "Museums with a mission. Education as it used to be", *Leeds Museums and Galleries Review* 3 (2000).
24. Barbara Woroncow, "From Rainy Day to Every Day: The Development of Museums, Galleries and Related Attractions in the Region", in *Managing Regional Tourism. A Case Study of Yorkshire, England*, ed. Thomas Rhodri (Ilkely: Great Northern Books, 2009), 33–34.
25. Peter Brears, "Temples of the Muses: the Yorkshire philosophical museums, 1820–1850", *Museum Journal* 84 (1984): 3–19.
26. Thomas Hughes, *Tom Brown's Schooldays* (London: Puffin Books, 1857).
27. Adrian Norris, "Vanishing herds—large mammals in museum collections?" *B.C.G. Newsletter* 4 (1988): 148–149.
28. The author personally visited one natural history department in one of Britain's largest cities in which the public display included a case of primates complete with strings of colored paper trimmings around the necks.
29. Eirik Granqvist, "Case study 4: a museum is dismantling its habitat dioramas", *ICOM Natural History Committee Newsletter* 17 (2006), accessed February 23, 2011, http://www.ggwinter.de/icom/nl_17.pdf.
30. Barry Lord, Gail Dexter and John Nicks, *The Cost of Collecting* (London: HMSO, 1989).
31. David T. D. Clarke, "Museum Disposals", in *The Value and Valuation of Natural Science Collections*, eds. John R. Nudds and Charles W. Pettitt (London: Geological Society, London, 1997), 104.
32. John R. Nudds and Charles W. Pettitt (eds.), *The Value and Valuation of Natural Science Collections* (London: Geological Society, London, 1997).
33. Adam Schoon, "Taxidermy—a resurgence of interest", *Bulletin of the Yorkshire Naturalists' Union* 52 (2009): 18–20.
34. Anonymous, *Collections Register Leeds City Museum (1822–1823)* (Leeds: Leeds Museum 1823).
35. Great Britain Department of Education and Science, *Provincial Museums and Galleries*, report of a committee appointed by the Paymaster General (London: Great Britain of Department Education and Science, 1973).
36. D. Andrew Roberts, *Planning the Documentation of Museum Collections* (London: The Museum Documentation Association, 1985).
37. Museums and Galleries Commission, *Standards 2 in the Museum Care of Biological Collections* (London: Museums and Galleries Commission, 1992).
38. Museums and Galleries Commission, *Standards 3 in the Museum Care of Geological Collections* (London: Museums and Galleries Commission, 1993).
39. "Collections Australia Network", accessed February 15, 2011, http://www.collectionsaustralia.net/.
40. "eHive", accessed February 15, 2011, http://ehive.com/.
41. "Smithsonian Institution Research Centers", accessed February 15, 2011, http://www.si.edu/research/online-collections-datasets.htm.
42. "FENSCORE", accessed February 15, 2011, http://fenscore.man.ac.uk.
43. The Conservative-Liberal Coalition Government elected in 2010 has decided to discontinue the MLA, as a cost-cutting measure.
44. Museums, Libraries and Archives Council, *Renaissance for the Regions: A New Vision for England's Museums in 2001* (London: Museums, Libraries and Archives Council, 2001).

45. John R. Nudds and Charles W. Pettitt (eds.), *The Value and Valuation of Natural Science Collections* (London: Geological Society, London, 1997).
46. Barry Lord, Gail Dexter and John Nicks, *The Cost of Collecting*.
47. In February 2011, the European Commission issued new official guidance stating that rhinoceros horn mounted on a plaque/shield or other type of base has not been sufficiently altered from its natural state to be included in the derogation for "worked" specimens and are not be considered to meet the antiques derogation or will be able to be able to be legally offered for sale. This definition has also been adopted in the UK.
48. A. Sailor, *The Echalax Bird Collection* (London: MPM Publishing, 2007).
49. See also the discussion by John Long, Chapter 3.
50. Pat A. Morris, *Rowland Ward, Taxidermist to the World* (London: MPM Publishing, 2003).
51. Pat A. Morris, *Edward Gerrard and Sons: A Taxidermy Memoir* (Lavenham: Lavenham Press, 2004).
52. Pat A. Morris, *The Kings Choice. George W. Quatremain Artist and Taxidermist* (London: MPM Publishing, 2006).
53. Pat A. Morris, *Van Ingen and Van Ingen: Artist in Taxidermy* (London: MPM Publishing, 2006).
54. Pat A. Morris, *Walter Potter and His Museum of Curious Taxidermy* (London: MPM Publishing, 2008).
55. *Ibid.*
56. Patrick Kelly, "Animal magic", *Museums Journal* December (2008): 30–33.
57. Axis is a publicly funded database of artists and their work. It has existed for many years in the UK. "Axis", accessed February 23, 2011, http://www.axisweb.org/. See also "Culture24", accessed February 23, 2011, http://www.culture24.org.uk/history+%26+heritage/time/georgian+and+victorian/art27392.
58. See also a discussion of Hirst by Eric Dorfman in Chapter 1.
59. Eirik Granqvist, "Working group on the art of taxidermy and its cultural heritage importance", *ICOM Natural History Committee Newsletter* 15 (2005), accessed February 23, 2011, http://www.ggwinter.de/icom/fnl.htm.
60. Kevin L. Campbell, Jason E. E. Roberts, Laura N. Watson, Jörg Stetefeld, Angela M. Sloan, Anthony V. Signore, Jesse W. Howatt, Jeremy R. H. Tame, Nadin Rohland, Tong-Jian Shen, Jeremy J Austin, Michael Hofreiter, Chien Ho, Roy E. Weber and Alan Cooper, "Substitutions in woolly mammoth hemoglobin confer biochemical properties adaptive for cold tolerance", *Nature Genetics* 42 (2010): 536–540.
61. Daniel G. Gibson, John I. Glass, Carole Lartigue, Vladimir N. Noskov, Ray-Yuan Chuang, Mikkel A. Algire, Gwynedd A. Benders, Michael G. Montague, Li Ma, Monzia M. Moodie, Chuck Merryman, Sanjay Vashee, Radha Krishnakumar, Nacyra Assad-Garcia, Cynthia Andrews-Pfannkoch, Evgeniya A. Denisova, Lei Young, Zhi-Qing Qi, Thomas H. Segall-Shapiro, Christopher H. Calvey, Prashanth P. Parmar, Clyde A. Hutchison III, Hamilton O. Smith and J. Craig Venter, "Creation of a bacterial cell controlled by a chemically synthesized genome", *Science* 329, no. 5987 (2010): 52–56.
62. Barbara Tellman, *Invasive Exotic Species in the Sonoran Region* (Phoenix: University of Arizona Press, 2002).

3 Case Studies of Intangible Natural Heritage from Museum Collections

John A. Long

INTRODUCTION

Intangible natural heritage can refer to environments, geographic places or a set of values embedded within museum artifacts or specimens. In the discussion of this book (Chapter 9), Eric Dorfman and Janet Carding describe intangible natural heritage as heritable environmental forces that create biological and geological entities, as well as the resulting interaction with human communities. A definition of intangible values by Allen Putney[1] refers to "that which enriches the intellectual, psychological, emotional, spiritual, cultural and /or creative aspects of human existence and well being". To refer to objects of natural history as steeped in values of intangible heritage brings in the notion of things as 'events' or an 'affecting presence'.[2] Taking this further, Kiresheblatt-Gimblett suggests that

> natural heritage, conceptualized in terms of ecology, environment and a systematic approach to a living entity, provides a model for thinking about intangible heritage as a totality, rather than as an inventory, and for calculating the intangible value of a living system, be it natural or cultural.[3]

Taking these concepts into account, certain natural history specimens in museums can also demonstrate high intangible natural heritage value because of their enhanced cultural significance, as distinct from their scientific face value as voucher specimens or samples from specific localities. The value in a museum natural history specimen is first and foremost as a representative reference object of a particular biological or palaeontological species or type of mineral, but often such objects have hidden layers of meaning that can be historical, cultural, artistic or sometimes even abstractly philosophical.

Applied to museum specimens, this can be manifest in many ways but is measured by the breadth or multi-layering of cultural, historic, emotional or artistic stories that add extra value to a specimen. Such enhanced values might have been acquired at the time of discovery, on its journey to a museum or since it was accessioned. In a world of diminishing natural habitats, ever more museums specimens are becoming significant as the last examples of extinct species (e.g. the Golden Toad *Bufo periglenes*, Passenger Pigeon *Ectopistes migratorius*, Tasmanian tiger *Thylacinus cynocephalus*). Such specimens may also contain significant cultural values to indigenous peoples, as well as other value in relation to art (e.g. if painted before extinction by a famous artist like Audubon or Gould), or with layers of embedded social and/or political history (e.g. the cases of the Western Australian state fossil emblem and the Cervantes egg). Such objects are carriers of intangible natural heritage and become highly valued, sometimes even iconic specimens in museum collections.

Biological specimens comprise the vast majority of museum specimens in any natural history museum, For example the Natural History Museum of Los Angeles County (NHMLAC) has an estimated collection size of 35 million objects of which 34.5 million are natural history objects, and around 30 million of those are marine zoological specimens. Although biological specimens have prime scientific value in being conserved, representative reference specimens (holotypes, paratypes etc.), to be studied or sampled for analysis by future researchers to answer specific questions, the accompanying data concerning who collected them, wherefrom and at what time, may empower the objects with additional levels of significance.

When an object has been handled or studied by a great name in biology, for instance Charles Darwin, it can confer to a particular specimen a historical and cultural celebrity value. In cases where a specimen has been put to a purpose outside its expected uses (e.g. for research, exhibition or public programs), it may also enhance its intangible value. For example, a specimen might be the key exhibit in a landmark legal case. Some museum objects have even been used as subjects for works of art or are highlighted on popular television shows. A high degree of media fame associated with a particular specimen, of course independent of its intrinsic scientific value, adds to its status in the public mind in arguably much the same way Hollywood celebrities gain status through increased media exposure.

The large *Tyrannosaurus rex* skeleton mounted in the entrance foyer at the NHMLAC, engaged in a combat pose with the skeleton of a *Triceratops* sp., might seem to be impressive enough from its dramatic posture and the fact that it incorporates real *T.rex* bones. However, on telling visitors that this is the same skeleton replicated for use in the final scene of the 1993 Spielberg movie *Jurassic Park,* in which the predatory velociraptors are hunting the humans and leap over the mounted *T. rex* skeleton, it takes on even greater status. *Jurassic Park* was the first blockbuster movie to show dinosaurs in lifelike computer-generated animation and to elevate the cult following of dinosaurs to a billion-dollar industry through the success of the movie and its associated merchandise.

This chapter documents and discusses a number of case studies of such natural history specimens in major museums that can be classified as items containing important intangible natural heritage, determining circumstances and mechanisms that are drivers to empower such specimens with increased significance. I have divided the case studies into biological, paleontological and mineralogical/petrological categories.

BIOLOGICAL CASE STUDIES

An Icon of Global Climate Change: The Golden Toad

A frog in a jar of alcohol in the basement of the NHMLAC is now a universal symbol of our fate as a species. The golden toad (*Bufo periglenes*; Figure 3.1) was abundant in a small area (10 km^2) on the Monteverde Cloud Forest high up on the mountain tops of Costa Rica when collections were first made by Dr Jay Mathers Savage, then of the University of Southern California, in May 1964. On discovery of the beautiful deep golden toads he quotes "my initial response when I saw them was one of disbelief and suspicion that someone had dipped the examples in enamel paint".[4]

Due to rising temperature, the mists that regularly shrouded the cloud forest in moisture have been slowly drying up. The natural habitat of these creatures was a 30-square-kilometer area of high altitude, cloud-covered

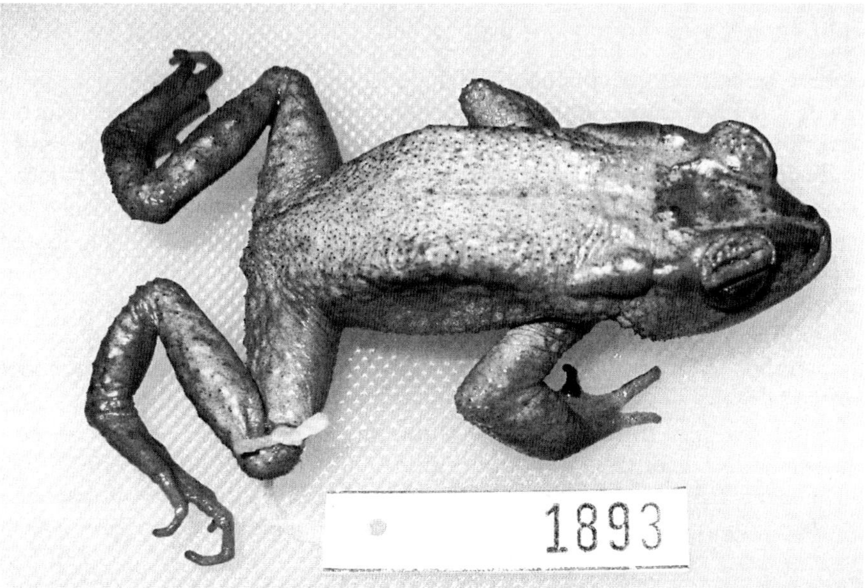

Figure 3.1 The Golden Toad of Costa Rica, an icon of global climate change, the first vertebrate species to go extinct entirely from shifts in climate represents intangible natural heritage values in a museum specimen. Holotype specimen, NHMLAC, collected by Jay Savage in 1964. Photo by Neftali Comacho, NHMLAC.

mist forests that regularly received moisture from the clouds formed by evaporation off the sea. A severe El Niño event in 1986–1987 saw a drying up of moisture in the region and the onset of decline of the toads. An American biologist conducted a survey in 1987 and counted some 1,500 toads. However the warm dry spring caused evaporation of pools before the tadpoles could mature. The following year only one toad was reported by that group, although another survey found eight males and two females. By 1989 only one solitary male was found.[5] The species was officially designated extinct by the International Union for the Conservation of Nature and Natural Resources (IUCN) in 2008. It is now recognized universally as perhaps the first vertebrate species to become extinct solely through the influence of anthropogenic climate change.[6] The area had been protected and was not influenced in any manner by the presence of humans, so the recent shift in climate (humidity in particular) is the clear primary driver of its extinction, with secondary causes being the Chytrid fungus that kills frogs, which had spread due to the changed weather conditions. Today, the type specimen of the golden toad of Costa Rica held in the NHMLAC, specimen number 1893, in the herpetological collections (Figure 3.1), conveys how natural history specimens tucked away in museum drawers can take on significant levels of intangible natural heritage, sometimes without the host museum even being aware of their new cultural significance.[7]

The Last Tasmanian Tiger and Other Historically Extinct Species

Other biological specimens carrying similar intangible natural heritage values include many recently extinct species captured in historical narratives, drawn by early natural history artists and steeped in local and/or indigenous folklore. A prime example is the Tasmanian tiger or thylacine of Australia (*Thylacinus cynocephalus;* specimens in many museums). The thylacine, a dog-like carnivorous marsupial closely related to the Tasmanian Devil, was widely hunted with bounty on it, not just by farmers to protect their sheep but also as a prized exchange to other museums in the early 20th century.[8] Eventually, the last captive animal died in Tasmania in September 1936, sending the species extinct. Some specimens were reportedly exported through the agency of Professor Theodore Thomson Flynn, the first biology professor at the University of Tasmania, who happened to be the father of the famous film screen star Errol Flynn. Flynn became a museum trustee in 1909 and curator in 1912, and it was reported that "other trustees thought he treated museum material as his own".[9] Such specimens, if they can be traced through associated notes about their source and export papers, would no doubt have additional intangible natural heritage value due to this celebrity connection.[10]

Many other examples of extinct biological icons exist in museum collections, including specimens of species that have died out recently enough to have featured in folklore and historical narratives, and some of which have been depicted by famous artists such as John James Audubon. The Steller's sea cow *Hydrodamalis gigas*, Passenger Pigeon *Ectopistes migratorius* and skeletons of the dodo *Raphus cucullatus* are prime examples.

Racehorses and Breeders: Biological Icons of 20th Century Culture

The mounted hide of the racehorse Phar Lap is arguably one of the best known displays in any Australian Museum; it is a true treasure of Museum Victoria's mammal collection. Similarly, his heart is one of the most sought-after displays at the National Museum of Australia, Canberra, as is his mounted skeleton at the Museum of New Zealand in Wellington. He is advertised on Museum Victoria's website as "Phar Lap, Australia's Wonder Horse" and was touted as a hero during the Great Depression of the 1930s. Born in 1926 in Timaru New Zealand, the horse arrived in Australia in 1928 in mangy condition. David J. Davis, the owner who purchased the horse was disappointed at seeing him and eventually made a deal with the horse trainer Harry Telford to keep him if Telford trained him for free, and he would keep one-third of the winnings. Phar Lap went on to win many major races, including the prestigious Melbourne Cup in 1930. In his four years of racing, he won 37 of his 51 races, and in 1931 he won 14 races consecutively. Shipped to Mexico in early 1932, he won the Agua Calinete Handicap, the largest purse ever for a U.S. race at that time. From there he was sent to California to compete but mysteriously died in San Francisco on April 5, 1932. The cause and mystery surrounding his death was as much the stuff of lore and controversy for many years as was his incredible race track record. In 2008 a detailed analysis of Phar Lap's hair by the Advanced Photon Source at Argonne National Laboratory, Chicago, lead by Ivan Kempson and Dermot Henry, identified a massive dose of arsenic, administered about 48 hours before his death, as the probable cause. This ingredient later showed up in the newly acquired Phar Lap's trainers tonic book (acquired by Museum Victoria's history department), and arsenic was an ingredient included in his regular tonics, leading to the conclusion that the horse was most likely accidentally poisoned to death.[11] Phar Lap has, since his death, been immortalized in film, by songs and books (including a recently published children's book by Museum Victoria) and knowledge of his story has been used as part of the test to become a nationalized Australian. Phar Lap's representative parts, now held in three major museums spanning the Tasman Sea, are wonderful examples of the richness of intangible natural heritage contained within zoological collections.

Another horse, represented by a skeleton in a U.S. museum nicely compliments the Phar Lap story and demonstrates how such skeletal mounts can hold intangible values for other reasons apart from racing fame. The skeleton of the Arabian horse, *Deyr*, was presented to the Los Angeles County Museum of History, Science and Art (now the NHMLAC) by Will K. Kellogg in 1927.[12] Kellogg wanted to have an Arabian horse skeleton displayed at the time to show the virtues of what was then considered to be a rare and exotic breed. *Deyr* was imported from what is now Syria by newspaper cartoonist, Homer Davenport in 1906. Davenport had become fascinated with the Arabian horse after seeing the ones that had been brought to the U.S. for the Chicago World's Fair in 1893. He contacted the man who had purchased them, Peter Bradley of Massachusetts, and eventually formed a partnership with him. He decided that he wanted to go to the desert himself to obtain the "real" horses directly from the Anezeh Bedouins. In 1906 he mounted an expedition to what is today Syria and Lebanon, using diplomatic assistance directly from his personal contact with President Theodore Roosevelt. He also received full support from the Sultan of the Ottoman Empire.[13] This was the first importation of horses directly from the desert to America and the chestnut colt, *Deyr,* was one of Davenport's horses.

Deyr is number 33 in the American Arabian horse registry and is considered to be one of the major foundation sires of the breed in this country, with thousands of horses having him in their pedigree. He is also one of the few individuals that remain of the pure desert Arabian horse. As such, he is an excellent example of a museum specimen in a zoological collection that transmits a rich cultural and, for horse breeders, a valuable genealogical history.

PALEONTOLOGICAL ICONS

Icons of Human Endeavor: Scott's Fossil Plants from Antarctica

Fossil plants dated at around 260 million years old that were collected on the fatal 1912 man-haul to the South Pole lead by Captain Robert Falcon Scott would later sow the first seeds of the idea of Gondwana, the great southern supercontinent. Scott and his men perished, but they dragged the geological specimens collected near the Beardmore Glacier on Thursday, February 8 on their way back from the Pole. As things grew worse, they still managed to haul the 16 kg of rock and fossil samples with them to their final depot, well after discarding all unnecessary equipment and supplies, knowing full well the extreme importance of what they carried. Scott even recorded in his diary on March 16 or 17 (he was unsure of the date) that they were feeling despondent about making it back alive, but still they hauled the rocks at Wilson's urging:

> *We are cold on the march now, and at all times except meals. Yesterday we had to lie up for a blizzard and today we move dreadfully slowly. We are at No. 14 pony camp, only two pony marches from One Ton Depot. We leave here our theodolite, a camera, and Oates sleeping bags. Diaries etc., and geological specimens carried at Wilson's special request, will be found with us on our sledge.*[14]

The Permian plant *Glossopteris* is today known to be a widespread form that occurred across the ancient supercontinent of Gondwana (Australia, India, South Africa and South America) back in the Permian Period (ca. 250–300 million years ago). The first discovery of this plant in Antarctica back in 1912 was seen to be of great scientific significance. Seward[15] identified *Glossopteris indica* from Scott's Antarctic samples, and this indicated that species of fossil trees found in the Permian of India were also present on the ancient Antarctic mainland, sowing the first scientific seeds of the concept of "Gondwana". These iconic fossils are now housed in the Natural History Museum, London. They symbolize fervor and endeavor in the heart of humanity's quest for knowledge against all odds, from the heroic age of Antarctic exploration, found in the same tent with the dead bodies of Scott, Wilson and Bowers and, as such, are truly prime examples of museum specimens rich in intangible values.

Super-Celebrity Status in a Museum Specimen: Darwin's Lost Fossil Barnacles

Charles Darwin was well known for his detailed research into extant and fossil barnacles, authoring a major monograph on the subject in two parts.[16] In 2008 fossil barnacles held in Museum Victoria in Melbourne, Australia, were discovered by the staff to be the very specimens described and figured by Charles Darwin and crucial to shaping his thinking when formulating his 1859 theory of evolution. Their story signifies how seemingly ordinary, unassuming objects in a museum cabinet can be elevated overnight to iconic specimens not only of immense scientific value but also carrying a legacy of great social and cultural significance.

How these specimens ended up in a museum in Australia is another story necessary to our general understanding of the fundamental growth of museum collections over the past two centuries and explains why just about any large museum collection from anywhere in the world has the potential to contain similar hidden treasures. In the early days of Museum Victoria, set up by its founding Director Dr Frederick McCoy, numerous natural history collection items were acquired through a program of acquisition designed to speed the growth of the museum's scientific reference collections. The private collection of fossils containing Darwin's barnacles was one that had been purchased from England about 10 years or so after Darwin's work on them had been completed. We can only assume that Darwin borrowed the specimens from a private collection to do his work and that at a later time the collection was sold to the museum without the barnacles receiving any special attention. It was only much later, in the 1980s, that the curators of invertebrate paleontology made the connection by matching the actual specimens with images form Darwin's 1851 monograph. Specimens showed the same physical characteristics, with identifying marks such as small breaks and minor damaged parts that matched the pen and ink illustrations in Darwin's work.

Today, these specimens are of immense interest to anyone who is shown them. Indeed there are few specimens in museums that can be attributed to being handled and studied by Charles Darwin himself in the formative years leading up to his major publication on evolution. Indeed, many specimens exist in museums around the world that were handled and studied by Charles Darwin and other famous early biologists, but most are accounted for. All of these specimens by nature hold intangible natural heritage values above their significance as voucher specimens.

Science and Political History: The Western Australian State Fossil Emblem

A 380-million-year-old fossil fish found in north Western Australia became the center of a political campaign that resulted in Australia's first state fossil emblem being declared in December 1995. It is one of the very few documented cases[17] showing how a fossil specimen in a museum was the central focus of a major political campaign involving some of the most notable political figures of the day. Today most states in the U.S. have their own fossil emblems, but this was the first one to be officially named for any Australian State (and remains the only one to date). Such specimens deepen public understanding about the natural environment, inviting us to add a fourth dimension—time—into our contemplations of the natural world,

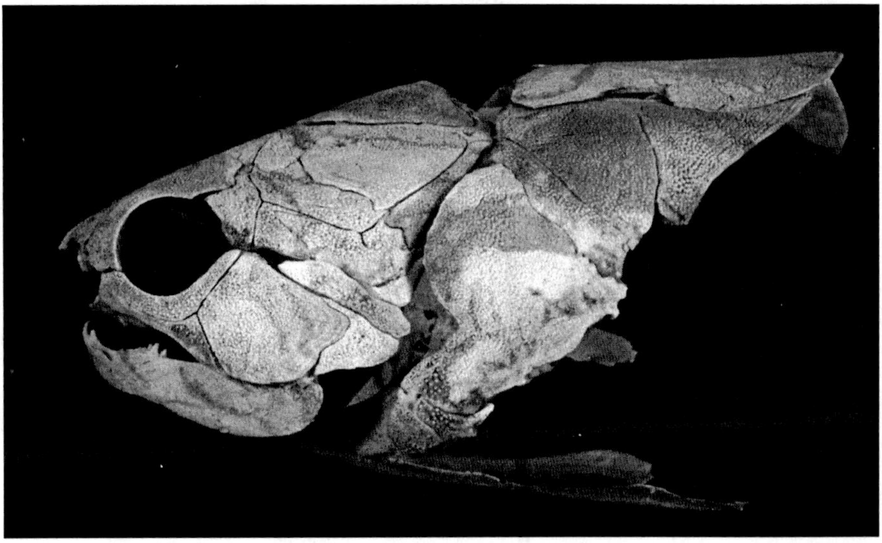

Figure 3.2 Mcnamaraspis kaprios, a 380-million-year-old fossil placoderm fish from the Gogo Formation, Western Australia (Western Australian Museum 86.9.676), is the official State Fossil Emblem of Western Australia and Australia's only proclaimed state fossil emblem. This specimen, the holotype, became the center of a political campaign in 1994–1995. Photo by the author.

and to comprehend that nature as more than just the configuration of animals plants we see today, but part of an ever-changing evolutionary process spanning millions of years.

The specimen designated as the official state fossil emblem of Western Australia is *Mcnamaraspis kaprios,* a placoderm fish from the Devonian Gogo site in Western Australia, collected by the author in 1986 and described and named 10 years later (Figure 3.2).[18] A year or so before this, children and teachers at Sutherland-Dianella Primary School began a campaign to lobby the Western Australian Government to establish a state fossil emblem. The Minister for the Arts at the time, Honorable Peter Foss, agreed, and there was a call for submissions from the public. Eventually, the Gogo fish fossil won the popular vote over other contenders and was officially proclaimed as the state fossil emblem on December 5, 1995, by the Honorable Major General Michael Jeffery, Governor of Western Australia (who eventually became the Governor General of Australia between 2003 to 2008).

The Cervantes Egg—Setting New Legal Precedents

A 32 cm long egg from an extinct giant elephant bird, uncovered in the sand dunes near Cervantes in Western Australia, became the subject of a media and legal battle between the Western Australian Museum and the family who found the object. The family brought the specimen in to the Western Australian Museum for identification in late 1991, then retrieved it the following day after it had been given a provisional identification as an Elephant Bird egg from Madagascar (*Aepyornis maximus*). It had most likely drifted on the currents and became washed up on the beach near Cervantes. Strangely, it was not the first case of such an egg to be found on a Western Australian beach, as an earlier example had been found in the southwest corner of the state, at the mouth of the Scott River, near Augusta.[19]

The issue that made this specimen famous was the legal tussle that ensued between the finders of the egg, who wanted to auction it off to the highest bidder, and the State of Western Australia, as the egg came from crown land. Under the then Crown Lands Act, the fossil legally belonged to the state, but the legal representatives of the egg's finder challenged this ruling due to the supposed high value of the specimen, based upon the precedent of high *ex gratia* payments that had been made to finders of archeological sites (e.g. historic shipwrecks) in past years. A court ruling eventually decreed that the egg was indeed Crown property and that the family was to hand it over to the Western Australian Museum. During this time the egg received widespread coverage on national television and in print media. At that point the family reburied the egg in the sand dunes and did not want to surrender it. Further advice from their legal counsel saw an agreement struck: the Western Australian Government would pay the family a $25,000 *ex gratia* payment for finding the egg and set up a fund-raising donation box for the family near the exhibited egg, so the public could make additional donations to reimburse them. The egg was then handed over to the Western Australian Museum, where it resides today.

The egg and the topic of who is the legal owner of such 'finders-keepers' property became the focus of a major legal study by a Monash University Law lecturer, Dr Joyce Tooher, which redefined ownership of personal cultural property.[20] This specimen thus gives an excellent example of the historical, legal, emotional and cultural stories embedded within one specimen now on display at the Western Australian Museum in Perth.

MINERALOGICAL AND PETROLOGICAL ICONS

The Hope Diamond: A Gem Steeped in Mystery and History

The Hope Diamond, currently on display in the Smithsonian Institution, Washington, D.C., and valued at approximately U.S.$200–$250 million, is the world's largest existing blue diamond. Found in India more than 400 years ago, French merchant Jean-Baptiste Tavernier purchased the gem about 1653 and brought it to Europe where it was sold to King Louis XIV of France for 220,000 livres. The 'Tavernier Blue,' as it was then known, was about 115 metric carats. In 1678, the French royal jeweler Sieur Pitau recut the gem into a more modern style, much akin to its modern shape by, slimming it down to 67.125 carats. It was then renamed "The French Blue" and handed down to the next king, Louis XV and, subsequently, Louis XVI. Later, when the royal family were interned in prison during the French Revolution, the jewels were stolen, and the diamond disappeared from history. It resurfaced in 1812, when the London-based diamond merchant Daniel Eliason figured it in a catalogue, its first official listing in a publication. In 1839 the gem appeared, now in a simpler setting, within a published catalogue of the collection of Henry Philip Hope. Hope died that year and, for 10 years, the family disputed ownership of the diamond. Henry Thomas Hope the eldest son inherited the diamond, and it was displayed in the Great Exhibition of 1851 in London and in the 1855 Paris exhibition. The gem was eventually handed down to Hope's wife, Anne Adele, who on dying left it to Lord Francis Hope in 1887. Due to accrued debts, Francis eventually sold the diamond to Adolph Weil, a London merchant for £29,000. Weil sold the stone to Simon Frankel in the U.S., who took it to New York. In 1908, he sold it to a Mr Solomon Habib, who let it go at auction in 1909. It was purchased by a Parisian jewel merchant for 400,000 francs and resold in 1910 to Pierre Cartier for 550,000 francs. The Hope Diamond then was acquired by U.S. socialite Evalyn Walsh McLean after Cartier set it in yet another style. On her death the trustees of her state sold the jewel to New York diamond merchant Harry Winston in 1949. Winston donated the stone to the Smithsonian Institution's National

Museum of Natural History on November 10, 1958, sending it by mail through the post in a brown paper bag.

The story of the diamond, now a mineralogical specimen in a museum, is undoubtedly steeped in rich cultural associations, not just through the many famous people with whom it came into contact but also through the lore and legends associated with it. 'The Curse' of the Hope Diamond was based on a rumor (unfounded) that it was first plucked from a Hindu idol and how some of its later owners had particularly bad luck. For example, King Louis XVI and his wife Queen Marie Antoinette both had worn the diamond on several occasions and eventually were both beheaded. Evalyn Walsh MacLean's first son died in a car crash at age 9, her daughter committed suicide at age 25 and her husband was declared clinically insane.[21]

Indeed, other famous gemstones in museum collections are associated with similar rich intangible natural heritage. Those that were once part of the royal family jewel collections are especially steeped in mythology. The fist-sized emerald Unguentarium now kept in the Vienna Natural History Museum collection is another such case. The stone, possibly the largest emerald known at the time, was first acquired by the Hapsburg ruler Rudolf II in the early 17th century, and purchased from the Spanish who acquired the emeralds from the Muzo mines in Colombia. On commission from Emperor Ferdinand III in 1641, the well known lapidary Dionysio Miseroni hollowed it out and created a vessel to store unguents. Even dramatically reduced, it still weighs a massive 2,860 carats. It is said that during the end of the Thirty Years' War in 1645, Ferdinand tried to sell it to raise money in Genoa. The story goes that the jewelers refused to value it and saw that it would be too difficult to find a buyer so were not interested in it. It was also reported that the Russians offered several chest of pearls for it, and the Grand Duke of Tuscany once offered three tons of gold for it.[22]

A Martian Controversy: The Allan Hills Meteorite

The Allan Hills meteorite (ALH 84001, Smithsonian Institution, Washington, D.C.) is a classic example of an iconic piece of rock steeped in intangible natural heritage. First, it fell from the skies (as it is a meteorite) and was found in Antarctica. Second, it has been analyzed as a Martian meteorite, formed by a large impact knocking of a part of the planet Mars and sending it into space to become a meteor. Third, and most significantly, it was announced in August 1996 by a team lead by Dr David McKay from the National Aeronautics and Space Administration (NASA) that they had found in it traces of what appeared to be organically formed molecules, this being touted as the first evidence of extraterrestrial life.[23] When the discovery was announced, there was major controversy, and many people assumed that the molecular fossils were good evidence of extraterrestrial life, making headlines around the world. The U.S. President Bill Clinton publically announced the discovery on television. Since then, other papers have criticized the discovery and proposed alternative views that the molecules could be natural inorganic structures. Nonetheless, a recent review of all the opinions and data on the sample concluded that "None of the original features supporting our hypothesis for ALH84001 has either been discredited or has been positively ascribed to non-biologic explanations".[24]

Meteorites invariably hold many levels of intangible natural heritage, some, for example, were used as early iron for primitive tool and weapon making, and others were revered as sacred objects because of their fall from the heavens. One in particular, though, served a very special purpose that helps philosophically to define our place in the universe. Canyon Diablo, was used by Clair Cameron Patterson in 1953 to determine the first accurate age of the Earth and solar system at 4.550 billion years (+/– 70 million years). The specimens come from near the world's largest known meteor crater, the Barringer Crater in Arizona, and pieces of it are now held in many of the world's major museums (e.g. Field Museum, Chicago, the California Academy of Sciences, San Francisco, and the NHMLAC, all in the U.S.; the Natural History Museum, Paris, France, and the Canterbury Museum, New Zealand).

ACKNOWLEDGMENTS

To Eric Dorfman for encouraging me to contribute this chapter; for providing information on subjects treated in this chapter, I thank Jim Dines, Cathy McNassor and Alyssa Morgan (NHMLAC); Dr Patrick Greene for helpful discussions; and Mr Neftali Comacho (NHMLAC) for the photo of the Golden Toad.

NOTES

1. Allen D. Putney, "Introduction: Perspectives on the Values of Protected Areas", in *The Full Value of Parks: From Economics to the Intangible*, eds. David Harman and Allen Putney (Lanham: Rowman and Littlefield, 2003), 3–13.
2. Robert Plant Armstrong, *The Powers of Presence: Consciousness, Myth, and Affecting Presence* (Philadelphia: University of Pennsylvania Press, 1981).
3. Barbara Kirshenblatt-Gimblett, "Intangible heritage as metacultural production", *Museum International* 56 (2004): 52–65.
4. Jay M. Savage, "An extraordinary new toad (*Bufo*) from Costa Rica", *Revista de Biologica Tropical* 14 (1966): 153–167.
5. Martha Crump, Frank Hensley and Kenneth Clark, "Apparent decline of the golden toad: underground or extinct?" *Copeia* (1992): 468–474.
6. Tim Flannery, *The Weather Makers: The History and Future Impact of Climate Change* (Melbourne: Text Publishing, 2005).
7. See also Chapter 2 by Adrian Norris.
8. David Owen, *Thylacine: The Tragic Tale of the Tasmanian Tiger* (Sydney: Allen and Unwin, 2003).
9. Kathryn Medlock, "Minutes of a Meeting of the Tasmanian Historical Research Association Held in the Royal Society Room, Tasmanian Museum

and Art Gallery on Tuesday, 14 July 2009 at 8pm", accessed February 13, 2011, http://www.thra.org.au/documents/General_meeting_minutes_14_July_2009.pdf.

10. Sleightholme and Ayliffe's 2005 international database for the Thylacine specimens in collections could be a helpful lead to trace such specimens. Stephen Sleightholme and Nicholas Ayliffe, *International Thylacine Specimen Database,* CD-ROM, Master Copy (London: Zoological Society, London, 2005).
11. Ivan Kempsonand and Dermot Henry, "Synchrotron radiation reveals arsenic poisoning and metabolism in hair: the case of Phar Lap", *Angewandte Chemie International Edition* 49 (2010): 4237–4240.
12. Specimen number A.1815, accessioned July 1927.
13. Homer Davenport, *My Quest of the Arab Horse* (Whitefish: Kessinger Publishing, 1909).
14. Robert Falcon Scott, *Scott's Last Expedition* (London: Smith, Elder, 1913).
15. Albert C. Seward, "Antarctic fossil plants", *British Museum (Natural History), British Antarctic (Terra Nova) Expedition (Geology)* 1 (1914): 1–49.
16. Charles Darwin, *A Monograph on the Fossil Lepadidae, or, Pedunculated Cirripedes of Great Britain* (London: The Palaeontological Society, 1851).
17. John A. Long, *Swimming in Stone—The Amazing Gogo Fossils of the Kimberley* (Fremantle: Fremantle Arts Centre Press, 2006).
18. John A. Long, "A new plourdosteid arthrodire from the Late Devonian Gogo Formation, Western Australia", *Paleontology* 38 (1995): 39–62.
19. John A. Long, Patricia Vickers-Rich, Karl Hirsch, Emily Bray and Claudio Tuniz, "The Cervantes egg: an early Malagasy tourist to Australia", *Records of the Western Australian Museum* 19 (1998): 39–46.
20. Joycey Tooher, "Jamie and the elephant egg", *Australian Property Law* 6 (1998): 117–143.
21. Richard Kurin, *Hope Diamond: The Legendary History of a Cursed Gem* (New York: HarperCollins, 2006).
22. For general information on the Austrian royal collection, see Alfred Auer, Margot Rauch. Veronika Sandbilcher and Katharina Seidl, *Die Entdeckung der Natur* (Innsbruck: Schloss Ambrass, 2007). Further information on the Unguentarium from "Emerald Unguentarium (Vessel) in the Imperial Treasury in Vienna", accessed February 23, 2011, http://www.internetstones.com/emerald-unguentarium-in-the-imperial-treasury-vienna-2860-carats-schatzkammer.html.
23. David S. McKay, Everett K. Gibson Jr., Kathie L. Thomas-Keprta, Hojatollah Vali, Christopher S. Romanek, Simon J. Clemett, Xavier D. F. Chillier, Claude R. Maechling and Richard N. Zare, "Search for past life on Mars: possible relic biogenic activity in martian meteorite ALH84001", *Science* 273, no. 5277 (1996): 924–930.
24. William Jeffs, "Study May Cast Doubt on Some 1996 Evidence of Past Life on Mars". Press Release #J04–025, (Washington, D.C.: Johnson Space Center, NASA, May 5, 2004), accessed February 23, 2011, http://www.nasa.gov/centers/johnson/news/releases/2004/J04–025.html.

4 'That Singular And Wonderful Quadruped'

The Kangaroo as Historical Intangible Natural Heritage in the Eighteenth Century

Markman Ellis

The kangaroo, a marsupial of the family *Macropodidae*, is endemic to the continent of Australia. The term kangaroo refers to several distinct species of the *Macropus* genus, including the red *M. rufus*, Eastern grey *M. giganteus* and Western Grey *M. fuliginosus*, although the term has been used historically for the entire family.[1] Kangaroo species were, since time out of mind, a valuable traditional resource for aboriginal Australians and were accorded significant meaning in mythological and totemic systems. Within European culture, and more especially the discourse of natural history, kangaroo were 'nondescript' (a species or genus not previously identified) before the first Captain Cook expedition to the South Pacific, in 1768–1772. Since that time, the kangaroo has become one of the durable symbols of Australia as nation and culture: stylized representations of the kangaroo are used on the nation's coat of arms, on currency and postage stamps and by numerous significant national institutions, such as sports teams, the armed forces and the airline Qantas, as well as in countless ephemeral iterations in popular and high culture. This essay explores how the kangaroo produced this meaning in European culture, tracing the history of several specimens in detail, from first contact and capture to display and classification, and examining the construction of the species as an emblem for a distinct Australian identity.

This essay studies the historical event of the encounter between science and the kangaroo in the late 18th century. It follows this encounter as a narrative, from description and capture in Australia, through the animal's inscription in print and visual media, to its eventual display in London, first stuffed and mounted, and subsequently in live menagerie displays. This narrative shows how the kangaroo was produced as an object of imperial knowledge in both metropolis and colony. The practice, discourse and discipline of natural history played an essential role in this narrative, legitimating documentary evidence of the encounter in the field and subsequently in the laboratory, museum and menagerie. Yet this narrative also records moments when the everyday otherness of the kangaroo as animal intrudes into its encounter with the human world. The essay therefore notices diverse forms of imaginative disruption and cognitive dislocation occasioned by the experience of the kangaroo: eruptions of non-scientific languages and practices, rhetorics of wonder and bewilderment, cooking and consumption, practices of field taxidermy and menagerie display, showmanship and heraldry. Intangible natural heritage should recognize the broad array of these efforts to make meaning out of the encounter with the kangaroo.

When the kangaroo first arrived in London in the late 18th century, the animal was constructed as an object of knowledge in popular culture as well as in the scientific discourse of natural history. Written and visual reports of the kangaroo, and later stuffed and live specimens, created a media controversy in Britain, focusing especially on the species' anomalous and wonderful characteristics. The encounter was mediated through a diverse range of intellectual practices and took place at diverse sociocultural levels. The intangible and creative process that produced –invented –the kangaroo as an object of natural history, perceived, understood and described the animal through a series of artistic and scientific practices, including field observation and collecting, menagerie display, mounting and taxidermy, anatomy, classification, culinary preparation, scientific writing, letter and diary writing, spectacle management, advertising copywriting, painting and printmaking. Through the process of inscription and debate, the kangaroo, an anomalous nondescript at once wonderful and bewildering, emerged in the London press as an important symbol for the penal colony at Botany Bay and later as a synecdoche for Australia.

In its focus on the late 18th century, this essay will provide an example, and outline a notion, of what I will call here *historical* intangible natural heritage. As Eric Dorfman and Janet Carding explain in the concluding discussion to this volume, the term 'intangible natural heritage' refers to 'the environmental forces that create biological and geological entities, the phenomena that these entities produce, and their interaction with humans and human communities.' The concept is developed from notions of intangible cultural heritage, as outlined in the United Nations Educational, Scientific and Cultural Organization (UNESCO) *Convention for the Safeguarding of Intangible Cultural Heritage* (2003), where the purpose was to identify, grant official recognition to and record instances of intangible cultural heritage, especially of practitioner communities in traditional cultures and folklore.[2] The intangible characteristics of heritage are probably always historically particular, and vestiges of deep historical particularity can be visible, howsoever partially occluded, in intangible heritage encountered in the present day. The conception of intangible heritage, as Frank Hassard has argued, is "essentially opposed to an established vision of heritage residing solely in the *materiality* of the past", although the system of recognition, recording, and conservation itself produces a material archive. 'Safeguarding,' as the convention envisions it, is necessarily tangible. As Hassard's analysis of the role of intangible cultural heritage in architectural restoration suggests, the relation between the intangible and the tangible is one of dialogue, in which one is reconciled to the other by a process of hermeneutic conversation. As he observes, research into intangible cultural heritage is a process, which uncovers "the methodological moment when the intangible 'adheres' itself to the tangible and becomes the historical document of the future".[3] For intangible natural heritage to become historical implies that those intangible qualities of practice and discourse enter tangibility as historical records and evidence. In the case of the kangaroo in the late 18th century, the historical context is centrally important: the evidence for the intangible interaction between biological entities and human communities described here is now only accessible through the historical archive. Newspapers, engravings, manuscripts, autograph letters, volumes of poetry and travel accounts, printed taxonomies, handbills and advertisements contribute the tangible but fragmentary archive of records that allows this intangible heritage to be recovered.

While the kangaroo narrative offered here is linear and teleological, it also gives evidence of complicated leaps of information between metropolis and colonial space. The first key locus for this is found in the practices and documentation of 'explorers' and colonists: initially by the Captain Cook expedition to New South Wales on his first voyage around the world in 1771 and then subsequently by the officers and convicts of the nascent penal colony in Eora country, known as Sydney Cove, Port Jackson, in the period 1788–1795. The second key locus is London (and more widely Europe) in the same period, especially in the early 1790s, when the kangaroo briefly became the focus of a celebrity media event. This history of the kangaroo as a public phenomenon is itself located at an early stage of an important transitional moment in the history of animal-human relations. An important, but broad-brush, theorization of this transition was offered by John Berger in his chapter 'Why Look at Animals?' (1977), in which he describes the physical and cultural "marginalization of animals" in modern urban life.[4] Where in pre-modern culture, animals were embedded in the everyday life of human societies, in the late 18th and early 19th century, animals begin what he calls their 'great disappearance.' In the countryside, the culture of agricultural improvement leads increasingly to the industrialization of animal husbandry; in the city, animals are more and more segregated into new categories of the pet, of vermin and the zoo. Ejected from everyday life, animals are transformed into spectacle, subject only to observation, cut off from their wildness and difference. But as Berger reminds us, and the evidence presented here reiterates, the encounter with a kangaroo in a menagerie, or with a preserved, stuffed and mounted specimen, is still an encounter with the animal's essentially other wildness, however much mediated by the conditions of its display in the museum, spectacle or zoo, situated in the midst of a densely populated, modern, urban human city.

CAPTURE

The first encounters between the science of natural history and the kangaroo were provided by ship-born expeditions in the 18th century. Although wallabies had been observed and recorded by Dutch sailors in the 17th century, the first extended description undertaken of the kangaroo was that of Joseph Banks on the eastern seaboard of Australia, on Captain James Cook's first voyage on the *Endeavour* in 1771. After a frustrating series of fleeting glimpses, such as the sighting of a 'quadruped about the size of a Rabbit' on May 1 in Botany Bay, the expedition finally secured a specimen when one was shot on July 14, 1771, near the Endeavour River in north Queensland. Banks's description, here revised by the official writer John Hawkesworth, clearly established the animal's anomalous and nondescript status.

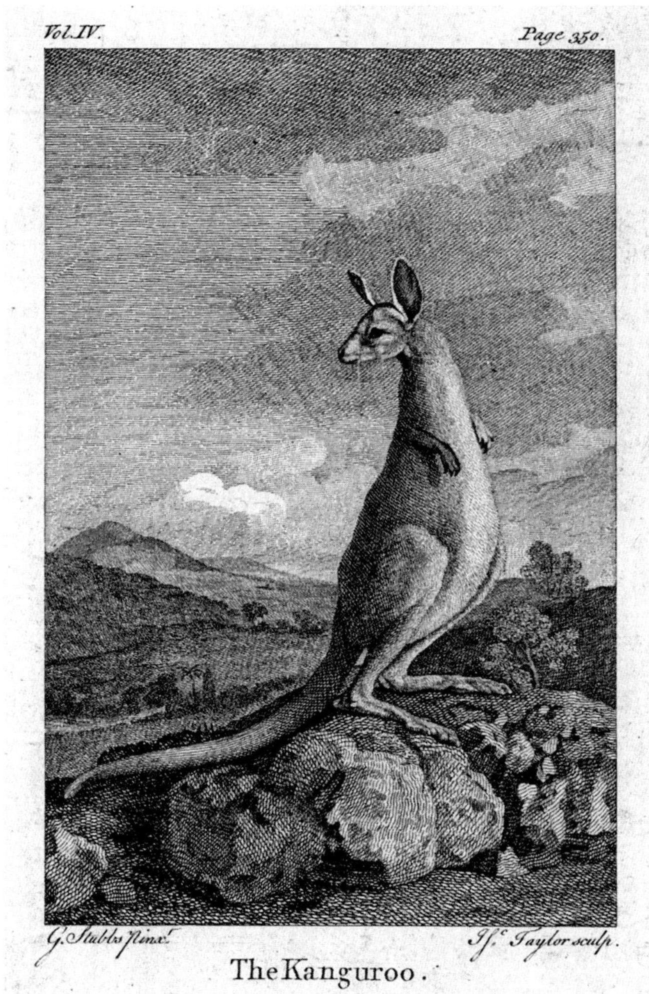

Figure 4.1 George Stubbs, engraved Isaac Taylor, "The Kanguroo", (London, n.d., 1774–1829) British Museum Department of Prints and Drawings: Location No.: British XVIIIc Unmounted Roy, © Trustees of the British Museum.

> *In form, it is most like the gerbua* [an archaic term for 'jerboa,' Family *Dipodidae*], *which it also resembles in motion . . . but it differs greatly in size, the gerbua not being larger than a common rat, and this animal, when full grown, being as big as a sheep: this individual was a young one, much under its full growth, weighing only thirty-eight pounds. The head, neck, and shoulders, are very small in proportion to the other parts of the body; the tail is nearly as long as the body, thick near the rump, and tapering towards the end: the fore-legs of this individual were only eight inches long, and the hind-legs two and twenty: its progress is by successive leaps or hops, of a great length, in an erect posture; the fore-legs are kept bent close to the breast, and seemed to be used only for digging: the skin is covered with a short fur, of a dark mouse or grey color, excepting the head and ears, which bear a slight resemblance to those of a hare. This animal is called by the natives* Kanguroo.[5]

According to the testimony of Banks and other members of the expedition, a description of the kangaroo using the available discursive machinery could only be achieved by a kind of analogical fragmentation: the kangaroo was, variously, like a rabbit, a deer, a hare, a sheep, a lamb and a jerboa. When the expedition returned to London in 1772, the kangaroo, one of the few substantial quadrupeds they had encountered, became one of the voyage's most celebrated discoveries. A preserved kangaroo skin was stuffed or inflated and painted by two artists, George Stubbs and James 'Athenian' Stuart.[6] The Stubbs painting was exhibited at the Royal Academy and subsequently engraved for Hawkesworth's official volumes of the expedition. The animal was shown at rest, looking back over its shoulder, as if pausing in flight so as to ascertain whether it was being followed. The curious historical intangible natural heritage of the kangaroo had begun (Figure 4.1).

After Cook's first expedition, no further contact was made between Britain and Australia until 1788, when Botany Bay, on the east coast of New Holland, as it was then known, was chosen as the location for a convict settlement for reception of prisoners condemned to transportation. The First Fleet of 11 ships left England on May 13, 1787, and arrived at Botany Bay on January 18, 1788, with 750 convict prisoners and 582 officers, guards and crew. The first news of the Botany Bay colony received in London arrived on March 26, 1789, when the *Prince of Wales*, a transport under the command of Captain Moore, docked at Falmouth, bringing 'an account of the fleet being all safe arrived at Botany Bay.'[7] She had sailed for England on July 15, 1788, with the *Borrowdale*, sailing via Rio de Janeiro.[8] An account of the expedition by the governor, Commodore Arthur Phillip, was immediately submitted to Thomas Townshend, Lord Sydney, at the Admiralty, where a copy was prepared for the King.[9] Other letters and accounts were the subject of news stories printed in numerous newspapers on March 26 and 28 (entitled variously 'Expedition to Botany Bay' and 'Accurate Particulars').

The London Chronicle (Thursday, March 26) reported the main news from the colony, such as its removal to Sydney Cove, Port Jackson, and the first preparations of the colony. But it also made some brief comments on the expedition's engagement with local inhabitants and natural history.

> *"The natives live chiefly on testaceous fish [shellfish], and the small quadruped which Cooke describes: the hind legs of which are much longer than the fore ones. The skins of several of these animals have been stuffed and brought to England. An attempt was made to bring some alive, but failed".*[10]

As this report suggests, the kangaroo was known primarily through the prism of the Cook voyage accounts, which had distinguished status in England. On Saturday, March 28, a fuller account was published, entitled 'Accurate Particulars', in the *Whitehall Evening Post* and other papers.[11] Under the heading 'Animals', the paper reported:

> *The kanguroo is as large as a sheep:–the head, neck, and shoulders are very small in proportion to the other parts of the body: the tail is long, but thick near the rump, and tapering towards the end: the fore legs in general measure only eight inches in length, and the hind legs twenty-two:–the progress is by successive leaps or hops of a considerable length, in an erect posture; the fore legs are kept bent close to the breast, and seemed to be of use only for digging: the skin is covered with short fur of a dark mouse or grey color, excepting the head or ears, which bears a slight resemblance to those of the hare. One of these animals, of uncommon magnitude, is on board the* Prince of Wales; *a live dog from New Holland is also brought to England; and besides these animals there are no other to be noticed except the Opossum and the Polecat.*

This account was widely repeated in newspapers, monthly periodicals[12] and pamphlet accounts, with little variation. It is of course an adequate but unenlightening description of the kangaroo (*M. giganteus*), making some attempt to adopt the descriptive technique and rational discourse of natural history. Some versions reiterate that the animal is a quadruped, one says it hops like a frog, and some instruct readers in an appropriate aesthetic language (that it is an 'uncommon curious' or 'remarkable' animal). The kangaroo is also subject to the practice of collecting: the newspaper notes that the skin of a large specimen had been returned on the *Prince of Wales* transport along with a live specimen of a dingo (*Canis lupus dingo*). A further account on March 30, also brought by the *Borrowdale*, mentioned that "there is not any animal in the country but the Kongaroo somewhat like a deer";[13] although other reports suggested the only "natural wonders from Port Jackson in New Holland are a black swan and a lobster, which was red when first caught".[14] While the reports from Botany Bay concentrated on the human history of the convict settlement, the natural history was also the object of extensive curiosity.

The Cook expedition had established the name 'kangaroo' for the genus, unaware that it comprised numerous species. The naturalist Sydney Parkinson had collected a vocabulary of 'native' words in the vicinity of the Endeavour River, including the word "Kangooroo" for "*The leaping quadruped*".[15] Hawkesworth's official description noted that the animal was "called by the natives *Kanguroo*",[16] although the expedition supported numerous different ways of spelling this word, including 'kanguru,' 'kangooroo,' 'kangouro' and 'ganguru'. Adopting the names used by indigenous peoples for natural history was an accepted practice of the period for nondescripts. Research by Raphael Cilento defends the name as a "reasonable transliteration" of the Guugu-Yimidhirr word *ganguru*, which signified one of the five kangaroo and wallaby species indigenous to the Endeavour River, namely the species now known as the Eastern grey kangaroo *M. giganteus*.[17] However, as the first fleet officers discovered, the word was not used by Dharawal-speaking Cadigal people who inhabited Eora country, known to the expedition as Botany Bay and Port Jackson.[18] The consensus among the officers in Sydney was that the kangaroo should be known by the transliteration 'pattagorang', from the Dharawal language of the Eora people. As Captain John Hunter explained, "The animal described in the voyage of the Endeavour, called the kangaroo, (but by the natives [of Port Jackson] patagarang) we found in great numbers".[19] Attempting to set a standard for accuracy, Governor Phillip described his specimens, in his correspondence with London, as the animal 'known in England by the name kangaroo'.[20] One observer who noted this confusion early was a midshipman on the *Sirius*, Newton Fowell, who reported to his father on July 12, 1788, that the 'Natives' around Sydney Cove referred to the first fleet's cattle as 'Kangooroo', implying that he believed them to have assumed that the word kangaroo was an English one meaning 'meat' or 'large quadruped'.[21] This misprision of names is celebrated in an apocryphal anecdote, in Beaglehole's version, that "an aboriginal, asked what it was called, replied in his own tongue, 'I don't understand'—which expression Banks in turn misunderstood as the name" kangaroo.[22]

The first kangaroo shot by the expedition was on Friday, February 8, 1788, about three weeks after the fleet's arrival. Hunting parties were common thereafter, as the meat was considered good eating.[23] The culinary preparation of natural history was both a necessity and a mode of observation, part of the process of turning raw information into cooked knowledge. Banks had eaten the carcass of the first kangaroo he caught in July 1770, commenting "The Beast which was killd yesterday was today Dressd for our dinners and provd excellent meat".[24] Food shortages in the first fleet settlement also meant that the kangaroo's status as big protein gave them high standing as a game animal. The surgeon John White reported that "the Flesh is not bad Eating something like Coarse mutton".[25] Captain John Hunter commented that "we ate the flesh with great relish, and I think it good mutton, although not so delicate as that which we sometimes find in Leadenhall-market".[26] Kangaroo hunting, using dogs and guns, was a practice that could support both gentlemanly game hunting, provision of food and natural history. The surgeon on the *Lady Penrhyn*, transport, reported that on Saturday, March 29, 1788, while out hunting with the steward from the *Golden Grove*, supply ship, shot a female Kangooroo, wt. a young one in its false belly, the young one I preserv'd in Spts. & the other he stuffed himself".[27] As these accounts signal, acquisition of the kangaroo specimens was dependent on the practice of hunting, an activity that could be coded as gentlemanly (undertaken on horseback with dogs and guns) or in a more menial subsistence mode of food provision. In hunting the kangaroo, the human hunter enters the animal's habitat attempting to understand and know it, in order to locate and kill it.[28] Hunting had intangible mythic and symbolic significance in 18th-century British culture, as it did in classical Greek and Roman culture, and, differently, among the Aboriginal peoples of Eora country.

Given the size of the first fleet expedition, it is surprising that no provision had been made among its contingent for science, despite the close engagement in its planning by Sir Joseph Banks, President of the Royal Society. But Banks had evidently made private arrangements with several officers on the expedition to collect specimens of interest to natural history. These arrangements were accorded high priority by some of the officers, especially as collecting natural history specimens was one of the few activities they could prosecute in Port Jackson that offered any chance of financial advantage or cultivating patronage. Not all such attempts were successful. The expedition's chaplain, the Rev Richard Johnson, was one who failed in his task, writing a long apology to Banks on July 8, 1788, explaining his "want of taste in things of this Nature—I do not profess myself a Naturalist, or a Botanist" and complaining that "things have been so new & strange; everything in such confusion, and having so many things to attend to, that I have not had that leisure or time I could wish for such Recreation".[29] But several of the officers, especially the surgeons, turned their hand to collecting and preparing curiosities for the London natural history market. This was a high-status pursuit in that it allowed them to open correspondence with scientifically minded gentlemen in London, in which their placement in the field could leverage their claims to expertise and politeness.

In the early years of the colony, natural history curiosities were the most notable—and almost the only—export of any value in the metropolis. Like hunting, collecting specimens was a field practice, especially in the case of specimens as lithe and fast as the kangaroo, but unlike hunting, it also required a sophisticated skill set to preserve the carcass or to take care of the live specimen. One of the first newspaper reports in London, entitled "Botany-Bay Expedition. Accurate Particulars", in *The Morning Herald* of Saturday March 28, 1789, made careful note of the expedition's natural history collecting:

> *Among the curiosities that will arrive, is a stuffed black swan, which is said to have been shot by Major Ross, the Lieutenant Governor: Several stuffed Kanguroos are also coming over, but it was impossible to take any alive of a growth to send. A young one, is however in Major Ross's possession, and intended as a present to his Majesty, whenever it can be conveyed safely to England.*[30]

Official letters from Governor Phillip himself to Sir Joseph Banks and Lord Sydney were brought in the Alexander transport, on its return to London, under the care of Lt John Shortland, arriving on May 13, 1789. With the letters, which gave a frank and guardedly optimistic account of the progress of the settlement, was a consignment of natural history curiosities. The receipt for them carefully itemized the contents and their method of preservation:

> *1 Can containing a Stuffed Kangurroo & several Skins*
> *1 Keg of Gum, 1 Keg containing a Young Kanguroo, & the Fox Bat in Spirits.*
> *1 Case Bottle of plants in Spirits of Wine*
> *4 Cases of Seeds & Plants*
> *1 Small Box of Gum*[.][31]

Phillip was an observant amateur zoologist: "In the Kanguroo which is stuffed, not to alarm Lady Banks, you will observe the feet to be different, to what is described in your Voyage. [. . .] The situation of the Penis in the Kanguroo, appears uncommon, between the Anus & the Testicles, & very close to the former".[32] Another keg containing a kangaroo and a kangaroo rat were received from Phillip in November 1789, brought on the *Fishburne* transport, Mr Browne master. Dennis Considen, assistant surgeon, sent a parcel of curiosities to Banks on board the *Golden Grove*, William Shairp master: "I have taken the Liberty of Sending home [. . .] a Kangarou skin all Properly prepared & stuff'd".[33] As Phillip noted somewhat forlornly, 'Kangaroo skins are the only Articles this Country furnishes at present".[34]

Besides preserved kangaroos, the penal colony had one more commodity to export: information. Accounts of the expedition, the colony, and its natural history, were a successful publishing venture in 1789. Botany Bay was a media event in London. Advertisements for Captain Watkin Tench's *A Narrative of the Expedition to Botany Bay*, printed by John Debrett, was advertised on Saturday, March 28,[35] and published on Saturday, April 4, 1789.[36] On Monday, March 30, 1789, Woodfall's *Diary* suggested the Botany Bay expedition was a suitable subject for a pantomime or ballet at the theaters.[37] The first pamphlet accounts of the expedition were published on Thursday, April 2, 1789: *The Authentic Journal of the Expedition under Commodore Phillips* [sic] *to Botany Bay*, published by Forster and sold also by John Stockdale (this contained also 'A Copy of a Letter from Captain Tench of the Marines').[38] Stockdale also advertised that "In press and speedily will be published" was the semi-official account of the expedition, *The Voyage of Commodore Phillips* [sic] *to Botany Bay, with an account of the Establishment of the Colonies of Port Jackson and Norfolk Island*, in "one large volume, Royal Octavo" on Saturday April 4, 1789[39] (although it was not published until November 15, 1789, when it appeared with an extensive subscriber list). "Gentlemen desirous of having fine Impressions are requested to send their names as early as possible to the Publishers".[40] Sir Joseph Banks, a subscriber, had his copy bound in kangaroo skin.[41]

Reports of the Phillip expedition to Botany Bay, especially the trials and hardships undergone by convicts and officers alike, had found a ready market in London in spring of 1789. The kangaroo quickly emerged as an important emblem for the expedition in the printed accounts. A satirist in *Stuart's Star* (March 31, 1789) wrote an ironic account of the sufferings of one transported convict, who suggested life was so difficult in the colony, that he would have preferred hanging at Newgate, complaining that "All day I nothing to do but hard work", with nothing to eat but "picking the tough tail of a Kangaroo".[42] In the first years of the settlement, the information it exported, including that of its natural history, played an important role in shaping the nascent penal colony's cultural and political meaning. Kangaroo and convicts produced the spectacle of New Holland.

STUFFED

Metropolitan kangaroo curiosity extended to public exhibitions of specimens brought back from the colony, at first stuffed, and later live. The first public display of a kangaroo, and hence of Botany Bay natural history, opened about a month after the return of the transports, on Tuesday, April 28, 1789, exhibited by an anonymous promoter. This was of course in advance of Banks receiving his 'official' specimens from Phillip on May 13, 1789, suggesting that commercial collecting operated more efficiently than the institutionally validated form directed by Banks and Phillip. Indeed Phillip later complained that he had been unable to prevent the commercial sale of a collection of natural history specimens to "people belonging to the Transports".[43] Competition for specimens was intense, both in Botany Bay and in London.

An advertisement in *The Morning Post and Daily Advertiser* on Tuesday, April 28, 1789, announced the exhibition of "A New Species of the Animal Kingdom", celebrated as a "Treat for the Lovers of Natural History, just arrived from Botany-Bay, and is to be seen at the Great Room in Panton Street, near the Haymarket, from Nine in the Morning till Seven in the Evening, that singular and wonderful Quadruped, called by the Natives Kanguroo".[44]

The room in which the kangaroo was exhibited, the Great Room in Panton Street, was a concert venue, built in 1696 by Thomas Hickford and used for performances by Handel in the mid-century but by the 1790s more often used for auctions and miscellaneous entertainments.[45] The area around this street, especially the Haymarket, was home to some of London's most well-known spectacles. At the price of one shilling, the kangaroo was positioned as an expensive and elite display: the same price as Boydell's Shakespeare Gallery, Germani's model of Rome, and Barker's Panorama of Edinburgh, also exhibited nearby that season. In further explanation, the advertisement noted that the species was

> *Described by the late justly-celebrated Naturalist Dr. Solander, and is the only one of its species ever brought to Europe.*
> *See Captain Cook's second Voyage, in his account of Endeavour River.*
> *N.B. The above was shot by Captain Reid, of the Boroughdale.*[46]

This description located the kangaroo within the culture of high-status natural history collecting and heroic imperial enterprise. Daniel Solander (1733–1782), the celebrated secretary and librarian of Sir Joseph Banks, had been at the center of London scientific circles: Cook was the greatest navigator and seamen of the period. The exhibitor was less experienced, for the advertisement makes several mistakes: it identifies the wrong Cook voyage (it was the first not second) and the description (it was by Banks not Solander, published in Hawkesworth, as above). The authenticity of the specimen was guaranteed by specifying the officer who shot it, "Captain Reid": the *Borrowdale*'s Master, Readthorn Hobson Reed. Visitors to the exhibition received a printed handbill detailing what they saw, a copy of which is preserved in *Lyson's Collectanea* (volume IV, part 1) in the British Library, dated by hand April 1789.[47] The handbill, also titled *A New Species of the Animal Kingdom, Or a Treat for the Lovers of Natural History*, supplemented the excitable rhetoric of the advertisement with a long quotation from Hawkesworth's account of Cook's voyage (*An Account of the Voyages [. . .] for Making Discoveries in the Southern Hemisphere* [1773]), again misattributed to Solander and the second voyage, which did not visit Australia.[48]

Nothing is known of the entrepreneur who mounted the Panton Street exhibition, nor of the specimen displayed there. By contrast with the field practice of hunting and collecting, the display of specimens involved the extraction of the animal from its natural habitat and its insertion into an urban one. This new human environment exerted a modifying force on the specimen, whether live or preserved. Preserving large quadruped specimens from decay was a technically challenging procedure in the period. Among the prescribed methods was immersion in alcohol in glass jars or barrels and forms of embalming in which the body cavity was filled with aromatic spices or a proprietary drying agent.[49] On Cook's voyage, Banks had brought his kangaroo specimens home as skins preserved in alcohol (brandy or spirits of wine). In Port Jackson, Phillip tried all available methods, preparing a stuffed specimen, cured skins and specimens preserved in spirits.[50] Reed may have followed suit. Although for research purposes, especially for systematists, the unmounted skins may have been of the most value, specimens were frequently mounted for public display by skilled technicians, reflecting the broader interests of the metropolitan market for natural history specimens. The term coined in 1820 for this craft was 'taxidermy.'[51]

The usual method used to prepare a large animal specimen for display in the 1790s was to preserve its skin so that it could be stuffed and mounted so as to appear in the living attitude. As the skeleton was usually discarded or removed for separate preservation, larger species like the kangaroo had to be mounted on an artificial framework of wood or wire, augmented with filling of chopped flax, raw cotton or similar materials. These techniques of display, however, brought with them further dangers, as the process itself might introduce inauthentic features.[52] The 'portrait' painting of the kangaroo by George Stubbs, exhibited at the London Society of Artists in 1773, appears to have been from a skin that had been overstuffed, giving it a more muscular upper torso than is to be expected.[53] It is clear that the Panton Street kangaroo was presented as a spectacle for popular entertainment, whatever else its status within contemporary zoological research. The identity of the technician or artist who mounted Reed's kangaroo specimen for the Panton Street exhibition is also not known, nor is the *mise-en-scene* of the exhibit. In this sense, the printed ephemera and newspaper advertisements that accompanied the exhibition, and especially the interpretative framework within natural history, have considerable importance in recording otherwise intangible aspects of the exhibition experience.

A few months after the Panton Street exhibition, in August 1789, a stuffed and mounted kangaroo was displayed at the Leverian Museum, perhaps using the same specimen. The Leverian's new kangaroo exhibit was announced in advertisements in the press.

> *Recently added to the Museum.—A Male Kanguroo of the largest species, being one of the most perfect specimens yet arrived from New South Wales. It stands upward of five feet from the ground, and exceeds in length by twelve inches that described by Mr. Pennant.*
>
> *This animal varies both in figure and properties from the usual course of nature, in the formation of quadrupeds. One principal variation is its moving in an erect posture, on its hind feet and its tail, by means of which, when pursued, it springs over bushes ten feet in height, and flies from its pursuer with astonishing swiftness, by successive leaps or bounds of as many yards in length.*[54]

The move from a disused concert hall to the more august context of a museum signals a distinct shift in the public display of the kangaroo from spectacle to science. The Leverian museum was an "admired Assemblage of the Productions of Nature and Art" collected by Sir Ashton Lever from the 1750s to his death in 1788. This collection was of considerable importance to contemporary natural history, as not only the largest but also the most systematically organized collection of specimens in Britain. Lever had exhibited it in Leicester House, a mansion on Leicester Fields, from 1773 to 1788, when it was purchased by a natural history entertainer, Parkinson, and transferred to new premises at the corner of Blackfriars Road and Albion Street. As Arthur MacGregor has argued:

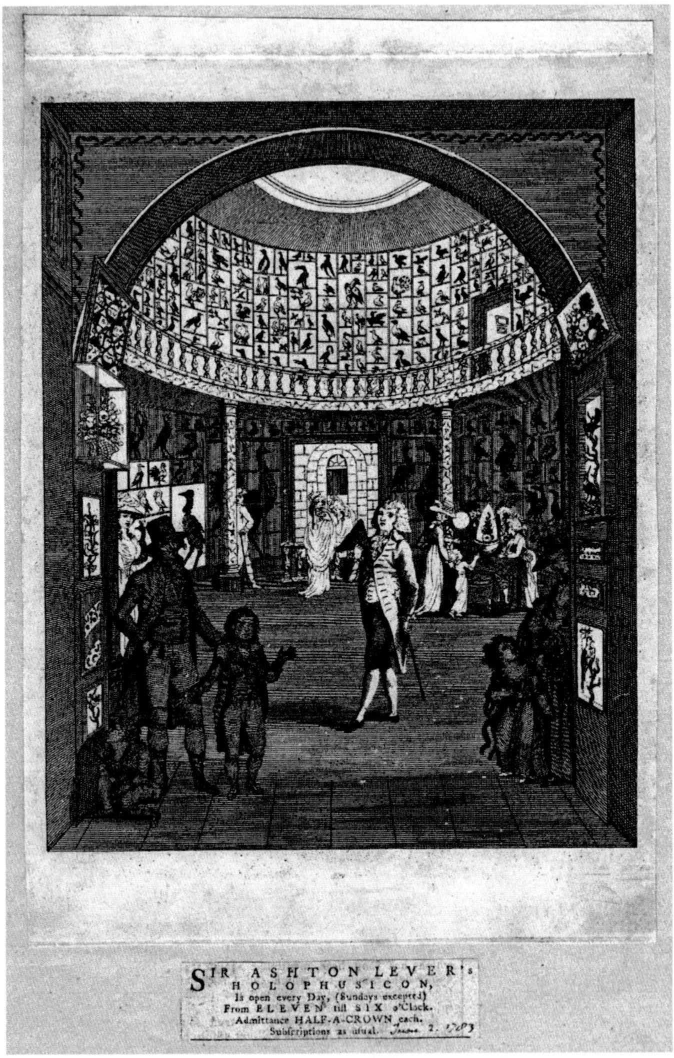

Figure 4.2 "Sir Ashton Lever's Holophusicon: View of the Grand Saloon and Gallery at the Leverian Museum" (1783), British Museum Department of Prints and Drawings, Location No.: Eu,EPH-AOA,B1.1 © Trustees of the British Museum.

> *In the matter of display, scientifically organized museums were initially resistant to exhibits involving groups of specimens arranged in picturesque tableaux: academic opinion preferred the displaying of birds and animals in a regular rather than an artistic manner, in standard poses on simple wooden bases against a plain background, in order to facilitate comparison and classification.*[55]

Surviving images of the Leverian Museum show this kind of systematic exhibition (Figure 4.2). "A Perspective View of the Grand Salon & Gallery of the Museum, late Sir Ashton Lever's, Surrey End of Black Fryer's Bridge", from *A Companion to the Museum* (1790) shows the interior of the museum as a grand rotunda lit from the top, with two circular galleries one above the other, displaying a regular patchwork of cases containing stuffed birds of diverse sizes.[56] In the middle of the room, an attendant carrying a tri-corn hat invites the viewer into the room: several groups already there include women, children and gentlemen. The space is open, egalitarian, heterosocial but also expensive, socially exclusive, scholarly and studious. The image further reinforces the primacy of museum display in the Leverian experience: what visitors see there is not a representation of the specimen's native location but the order of the museum, the typology of cases, frames and vitrines and the taxonomy of the species. The Leverian's advertisement in *The World* located the kangaroo within a natural history context, identifying it through Pennant's description from the second edition of his *History of Quadrupeds* (1781).[57] The description of its mode of locomotion suggests, however, elements of exaggeration not unlike that of the romance genre, both in the tripodic maneuvers of its feet and tail, and the tremendous scale of its leaps and bounds. Although the Leverian Museum identified itself as a space of science, these confusions and interpolations reallocate science as a form of spectacle.

The official institutions of London science also received kangaroo specimens, when a parcel of natural history curiosities from Governor Phillip arrived in London on May 13, 1789, on the *Alexander*. Although already familiar with the species himself, Banks recognized the kangaroo's value as a spectacular and modish scientific curiosity. He was instrumental in disseminating kangaroo specimens to natural history collections on the Continent and in the provinces. Banks had established a correspondence with a young French naturalist, Pierre Marie Auguste Broussonet, ever since he had visited Soho Square in 1780. Banks had promised Broussonet a kangaroo for the *Académie des Sciences*. On April 23, 1789, after the return of the transports, and presumably in response to the newspaper accounts of the kangaroo exhibition, Broussonet wrote again, pressing his claim: Banks replied on May 1 that "I shall remember you when I get a Kangouro[:] the Ship which Carries the Curiosities destind for me is not arrived but if it does it will bring one which is yours".[58] In June, Banks was able to send a specimen: it arrived in Dieppe on July 1, and in Paris on July 29, 1789.[59] A kangaroo specimen, stuffed and mounted for public display, was also exhibited in Edinburgh in the museum of natural history of Alexander Weir, a painter and taxidermist, at No. 31 South Bridge-Street. In January 1790, Weir advertised that among some recent additions was "that extraordinary Quadruped called The Cunquroo, from Botany Bay, In the most perfect state of preservation, being the first that was ever brought into Britain, which was presented to the Museum by a Lady of the First Distinction in this country".[60] Stuffed and mounted kangaroo specimens were constructed both as public spectacles, displayed in commercial exhibitions, and as curiosities of science, displayed in museums of natural history. The vogue, if it can be so called, for these specimens lasted a year or so: by 1791, they were supplanted in the metropolitan imagination by live kangaroo displays.

LIVE

The first live kangaroo specimens to be successfully returned to London were brought by the returning transports of the notorious Second Fleet, on which many convicts perished through disease and neglect. Attempts to bring live specimens home had been made for some years. Arthur Phillip noted in a letter to Banks in July 1788 that he had in his possession a juvenile kangaroo, captured live when its mother was shot.[61] "When young they are very pleasing, a few days makes them as tame as a Cat & they will follow the person that feeds them like a Dog. [. . .] But none have lived more than three weeks".[62] Keeping the specimens alive on the long trip home was also difficult. When the *Alexander*, a First Fleet transport, set off for its disastrous return journey on July 14, 1788, it left carrying a live juvenile kangaroo. William Bradley, a lieutenant on HMS *Sirius*, observed in his journal on July 11, 1788, that "One of the People killed a Male & Female Kanguroo & took a young one alive, which was sent aboard the Alexander Transport to go to England".[63] The animal, and many of the crew, did not survive the journey.

The convoy of Second Fleet transports and supply ships left Sydney in 1790, returning via Canton in China to take on a cargo of tea. The first to arrive in London was the *Surprize*, under Captain Anstis, which arrived on September 6, 1791.[64] Alongside the Chinese tea, the *Surprize* brought a collection of curiosities, which were auctioned at East India House in mid-October. "The curiosities from Botany Bay, by the *Surprize*, sold extremely well last week, at the East India House: a lot, consisting of three Kongarus, an Oppossum, and a Wolf, all in excellent preservation, were bought in by Mr. Walker, probably for Sir Joseph Banks".[65]

The trade in curiosities—including these preserved kangaroos—was one of the few profitable activities for the officers in the penal colony. More excitement was caused by the arrival of another store ship, the *Justinian*, which returned from Sydney, via Canton with a cargo of tea, arriving in Gravesend on Tuesday October 8, 1791, accompanied by the *Maitland*.[66]

As the *Evening Mail* reported, under a byline dated October 22: "The Justinian has brought a live Kongaru from Botany Bay—which is valued at the moderate sum of five hundred guineas –so that it will not very soon Jump from its owner!"[67]

The live kangaroo was a spectacle of the highest order: the sum of 500 guineas (£550) was an enormous sum (equivalent to more than half a million pounds now). Little is known about the master or crew of the *Justinian*, and the historical record does not indicate who owned this animal, nor to whom it was sent or sold.[68] But the excitement caused by its arrival was extensive.

From December 1791 live kangaroo exhibitions became one of the sensations of the London season. A newspaper, *The Morning Chronicle* (December 6, 1791), reported the exhibition as a newsworthy event:

> *There certainly was never seen in this country (or any other country in Europe) such a prodigious creature of the wild animal species, as that wonderful, animated, and uncommonly astonishing animal, the Kangaroo, now exhibiting at No. 31, Top of the Haymarket. The concourse of people of all ranks, and of both sexes, that daily resort to see it, is a fair instance of the truth of this assertion, and so captivated are they by the attractive powers of this so little known strange phenomenon, that many seem to leave its present place of abode with great reluctance, though having been minutely examining its beauties, tameness, &c. &c. for perhaps an hour or more already. Pity it is, that for the Convenience of the Public in general, the exhibition of it is not continued till nine o'clock, instead of seven, as great numbers in business are deprived of the pleasure of seeing it, owing to the Exhibition closing at so early an hour.*[69]

The story—probably written by the exhibitor—guides the reader, and the viewer, in the appropriate response to the animal, that of wonder and astonishment. The animal's curious oddity, its astonishing peculiarity, drives a numerous crowd to view it every day and causes them to remain a long time in viewing it. The location of the exhibition was, like the stuffed specimens, within the area of London known for the display of spectacles, such as the panorama and the theaters, and was just around the corner from the Great Room in Panton Street. The exhibition space at the Top of the Haymarket was located above the shop of Mr Becket's the Trunkmaker. It was a place known for displays of exotic animals (such as three gibbon *Hylobates* sp. in 1787), and for shows of magic and illusion.[70] Although the entertainment offered there was too expensive to be genuinely plebeian, it was nonetheless pitched at a popular audience hungry for the spectacular and the wonderful. The locality was decidedly not one dedicated to scholarship and scientific study. In the center of the emergent entertainment district of London's West End, this area was densely populated and intensely urban: a far cry from the kangaroo's indigenous habitat.

Advertisements for the animal ran in newspapers from early January, promising visitors an encounter with "The Wonderful Kanguroo, from Botany Bay", for the price of a shilling entry. The advertisement again records the specimen as a wonder: "to enumerate its extraordinary qualities would far exceed the common limits of a public notice". Its appeal is broad, both to men of science, and to the general public:

> *Let it suffice to observe, that the Public in general are pleased, and bestow their plaudits; the ingenious are delighted; the virtuoso, and connoisseur, are taught to admire! impressing the beholder, with wonder and astonishment, at the sight of this unparalleled animal from the Southern Hemisphere, that almost surpasses belief; therefore ocular demonstration will exceed what words can describe, or pencil delineate.*[71]

Daniel Lyson's collection of printed ephemera includes a handbill version of this advertisement, dated by its collector as 1791. The handbill also reiterates that the kangaroo's appeal is wider than the men of science:

> *It is not easy to describe that peculiarity of attitude, and uncommon proportion of parts, which so strikingly distinguish the Kanguroo from all other quadrupeds, and it may be presumed, that few who possess a taste for science, or a laudable curiosity of inspecting the wonders of nature, will omit embracing the only opportunity hitherto offered in Europe, of viewing this singular native of the Southern Hemisphere, in its natural state of vigor and activity.*[72]

The nature of the spectacle, the handbill explains, is the evidence imparted by the animals as a wonder of nature. The appeal to the language of natural history ("quadruped", "curiosity") locates the spectacle within the high status culture of the virtuosi, despite its context in the vulgar culture of menagerie spectacle.

Further live kangaroo specimens were imported from New South Wales in the following year. Governor Phillip sent some live specimens in late 1792 to the minister in charge of the colony, William Wyndham, Lord Grenville (Home Secretary 1789–1791, Foreign Secretary 1791–1793), believing them to be the first to arrive in London. Phillip wrote to Wyndham that an animal "which is known in England by the name of kangaroo" was to be sent with in HMS *Supply*, brig, when it returned to London, intending that it should be sent to the king. 'I have taken this liberty, as it is not known that any animal of the kind has hitherto been seen in England.'[73] A few weeks later, Phillip sent another of the 'uncommon' animals 'known in England by the name kangaroo' in the HMS. *Gorgon*, John Parker master, destined for the King.[74] Mary Ann Parker, the wife of the *Gorgon's* master, commented that the ship was "now crowded with Kangaroos, Opossums, and every curiosity which that country produced. The quarter-deck was occupied with shrubs and plants, whilst the Cabin was hung around with skins of animals".[75] When the *Supply* arrived in Plymouth on April 21, 1792, newspapers noted that "Several kangaroos were brought from Botany Bay, but only three remain alive on board at present, the others having been made use of on the voyage", presumably for food.[76] When the *Gorgon* arrived back in London in June 1792, John Parker wrote to Sir Joseph Banks about the consignment of natural history specimens for him. The plants, he said, should be consigned to a gardener sent from Kew. In addition, he wrote, "There is on board a Kangaroo, which I believe is consigned to you, and I believe it is for their Majesty's but I make no doubt but Governor Phillip had advised you of it".[77] When Phillips himself returned in May 1793 on the *Atlantic* transport, he brought two Aboriginal men, alongside "four kangaroos, lively and healthy, and some other animals peculiar to that country", described as "non-descripts".[78] Two of these kangaroos were presented by Phillip to his patron, Sir Joseph Banks, who in turn presented them to the King. As newspapers commented "They are now at Kew; their food is Indian corn".[79]

This official exchange of live kangaroos, as gifts-in-kind, developed alongside a burgeoning commercial system. Of the four kangaroos brought back by Phillips on the *Atlantic*, one, "a most beautiful Female Kanguroo", and two of the nondescripts, were exhibited commercially.[80] Two kangaroos seized by the Custom-House Officers in May 1792 were kept in the Tower of London until their ownership could be determined.[81] Commercial menageries competed for the public's appetite for live kangaroos in London. The first to display live specimens was Pidcock's Menagerie in December 1792, initially at the Lyceum in the Strand, a large exhibition space built in 1772 by the Society of Artists but used for a wide variety of public displays. George Pidcock advertised "the following wonderful productions of the Creation; amongst which is that singular animal the Kangaroo from Botany Bay, the only one ever brought alive to Europe; it distinguishes itself in shape, make, and true symmetry of parts quite different from all other quadrupeds &c."[82]

At Pidcock's, the kangaroo was exhibited alongside a wide range of exotic species, including zebra, rhinoceros, tiger, leopard, wolf, six macaws and a two-headed calf. As this range suggests, Pidcock was an energetic promoter of zoological spectacles, many of which attracted scientific interest but which were exhibited for primarily commercial purposes. As his newspaper advertisements proclaimed "Remember, they are all alive".[83] In January 1793 Pidcock purchased an existing menagerie of "Foreign Birds and Animals" (known as Clark's), including the lease on the space in which it had been exhibited, in the Great Room at Exeter Change in the Strand, and from February advertised his animals, including the kangaroo, as available to view at that address, admittance one shilling.[84] The kangaroo was now a centerpiece of what Pidcock called the "Grand Menagerie of 'Living

Rarities'", "the largest Collection of the Animal and Feathered Creation ever Exhibited to the Public".[85]

Pidcock's menagerie was one of the foremost locations for viewing exotic natural history in late 18th-century London. His menagerie was skilled at sourcing live specimens and presumably was willing to pay top prices. Its modes of display and advertisement suggest an uneasy mixture of scientific curiosity and vulgar public spectacle. Advertisements in late 1793 noted that the display included "two beautiful Kangaroos, Male and Female, from Botany Bay", as well as the "most wonderful Phænomenon in Nature", a cow with "Two Heads, Four Horns, Four Eyes, Four Ears, Four Nostrils, &c.," alongside a replica of "The Guillotine", the then newsworthy instrument of state execution in France during the Reign of Terror.[86] Visual evidence suggests that the Pidcock's menagerie in Exeter Change presented animals in cages stacked three deep in rows, one above the other, arranged around a central courtyard.[87] As recent zoo theory has suggested, the menagerie forms part of a culture of natural history display that is concomitant with early zoological gardens (such as the *Jardin des Plantes*, founded in 1793, or the London Zoological Society in 1828).[88] Berger argues that "Public zoos came into existence at the beginning of the period which was to see the disappearance of animals from daily life. The zoo to which people go to meet animals, to observe them, to see them, is, in fact, a monument to the impossibility of such encounters".[89] As in a zoo, the animal in a menagerie is ostensibly what there is to be seen. Yet what also visible, but overlooked, is the *mis-en-scene*: barriers (bars, shackles, muzzles), props (faux terrain, painted backdrops, artificial flora), isolation (each species exhibited separately) and temporal discipline (cleaning, feeding, public hours). As spectacle, the zoo assumes that the observer will overlook the surroundings and 'see' the animal: the actual subject of the display.

Live kangaroos were also to be encountered in aristocratic situations. The specimens at Kew given to the King by Banks in 1792 prospered, and a herd (a 'mob') was subsequently established in the royal menagerie at Richmond Park in 1793. By 1795, this had grown to number at least six, with "two or three of the females" having bred successfully in Britain.[90] Lord Grenville was sent several kangaroos by Governor Phillip and may have had them at his country house near Burnham in Buckinghamshire.[91] As with the stuffed and mounted specimens, Banks was recognized on the continent as the leading agent for live kangaroos. In July 1796, Banks was approached by a Florentine natural philosopher, Giovanni Fabbroni (1752–1822), about purchasing a live kangaroo for exhibition in the Grand Duchy of Tuscany. Banks replied that he knew of no "good Specimen of Macropus Giganteus" that was "upon Sale", but suggested that if the Grand Duke were apply to the King, then "I have no doubt he will Easily Obtain a male & a Female alive as they breed fast in the Royal Menagerie".[92] As well as Richmond Park, the King's kangaroos were later established in Windsor Great Park, where they were noted in great abundance in 1828, displayed in "spacious paddocks to range in", with "spreading trees to shade them from the noon-day sun".[93] William Bingley, in *Animal Biography* (1803), considered that kangaroos were "in a great degree naturalized in England",[94] and indeed colonies of wild wallaby still survive in small numbers in Britain.[95] William Beckford, in his novel *Azemia* (1798) suggested satirically that the kangaroo had become an "article of British luxury".[96] Kangaroos were exhibited subsequently in the grounds of the Prince of Wales' Pavilion at Brighton, and, by the Zoological Society of London, in Regents Park Zoo on its foundation in 1828. Displayed in the official and royal gardens, the kangaroo articulated an intangible imperial heritage, forging a metaphorical link in natural history between metropolitan Britain and its Australian colony.

CLASSIFICATION

More orthodox scientific work on the kangaroo proceeded alongside their exhibition to the public. Although the specimens brought to London by Cook's expedition were available for study, Hawkesworth's written account, a composite of observations by Banks, Cook and Solander, was the established primary authority, supplemented by the Stubbs engraving. Classifying naturalists relied on this description, rather than the preserved skin and skull brought home by the expedition, to locate the animal within the typology of quadrupeds, and thereafter, name it appropriately.[97] A series of natural historians, working within the Linnaean system, offered classifications: *Mus canguru* (Statius Müller, 1776); *Jerbua gigantea* (Eberhard Zimmermann, 1777); *Didelphus gigantea* (Peter Simon Pallas, 1777), *Jaculus giganteus* (Johann Christian Polycarp Erxleben, 1777) and *Yerbua banksii gigantea novae hollandiae* (Johann Reinhold Forster, 1777).[98] This plethora of names represented a very real confusion over the animal. In the Linnaean system of binomial nomenclature, names indicate order: "For a single genus, a single name".[99] Given this confusing debate over nomenclature, the holotype of the kangaroo remained a subject of debate, even among contemporaries. Phillip declared that the "first author who gave a scientific description of the kangaroo" was Thomas Pennant (1726–1798), in his *History of Quadrupeds* (1781), the standard work on mammalian zoology in English during the late 18th century.[100] Pennant described the kangaroo among the genus *Opossum*, although he had no knowledge of the kangaroo's marsupial status. He concludes his long description of the physical form and manners of the species with the comment "This is a very anomalous animal: but it has more relation to this genus than any other; and in form of its legs comes very near to the preceding species", the *Javan* opossum.[101]

Unfortunately, all three specimens returned by the Cook expedition were male, and key facts relevant to their classification remained obscure, primarily their marsupial mode of reproduction.[102] Information returned by Phillip's First Fleet expedition gave information about the female pouch, and its role in the gestation of the young: "The pouch of the female, in which the young are nursed, is thought to connect it rather with the opossum tribe".[103] But the significance of this insight on the animal's classification was largely lost on the colony's officers. Classification was clarified, finally, in 1790 by George Shaw (1751–1813), although even he despaired at the confusion surrounding the animal, complaining "The Kanguroo is a very anomalous and extraordinary quadruped". Shaw was a founding member of the Linnaean Society in 1788, elected a Fellow of the Royal Society in 1789 and appointed assistant keeper of the natural history section of the British Museum in 1791. His first scientific description of a kangaroo was published in June 1790 in his serial *Naturalist's Miscellany,* in Latin and English, illustrated with a plate—a description later supplanted by another in *Musei Leveriani Explicatio* (1796) and *Cimelia Physica* (1796), illustrated with plates by John F. Miller.[104] As he observed in 1796, "The kangaroo, the most extraordinary and striking animal which the Southern Hemisphere had yet exhibited to our view, may admit of being arranged differently by systematic naturalists".[105] The anomalous kangaroo expertly probes the different classification systems adopted by the noted authorities of natural history. The incipient contest between various artificial systems of classification, based on particular features such as teeth or reproductive organs, is encapsulated in Shaw's deft summary of the kangaroo's contraindicated classification.

The surgeon-anatomist Everard Home (1756–1832) undertook the most extended enquiries into marsupial generation, determining that the embryo is conveyed (by means he could not clarify) into the external pouch at a very early stage of its development, where it was nourished at two mammae. Homemade use of specimens collected for his brother-in-law, the surgeon and collector John Hunter (1728–1793), who had commissioned two officers traveling to Port Jackson (Captain Paterson and Mr Lang, surgeon) "to procure the female organs of the kanguroo under all the different circumstances in which they occurred and send them to England in spirits, that he might be enabled to prosecute his enquiry".[106] The "preparations" sent by the field researchers, together with observations made in Port Jackson by Dennis Considen, assistant surgeon, (*d.* 1815) provided Home with the evidence to determine the mode of the generation in the kangaroo. Home also observed the kangaroo in the royal menagerie at Richmond, which had two or three females that had bred in England, but without learning anything relevant. The "several preparations of the uterus in different states" were preserved in Hunter's collection, which was sold in 1799 to the College of Surgeons to form the Hunterian Museum.[107] Home's paper was read at the Royal Society on March 5, 1795 and published in the *Philosophical Transactions* in late 1795, with a large number of explanatory plates. As this debate indicates, kangaroo research was located in overlapping zones of discourse and practice: field observation, taxidermic preparation, anatomy theater, dissecting laboratory, menagerie, lecture room and periodical publication. As Home concluded

> *the peculiarities in the structure of female organs of the kangaroo, it is evident they must, in their mode of generation, materially differ from other quadrupeds. [. . .] The mode of generation just described will be found, in this chain of gradations of nature, to form a link between animals whose young are nourished by means of a connection with the uterus [mammals], and those that are nourished independent of it [oviparous, or egg laying].*[108]

Home's conclusion further underlined the anomalous status of the genus.

EMBLEM

In its transition from the spectacular displays of Haymarket and the newspaper advertisements, to the royal menagerie at Richmond and the pages of the *Philosophical Transactions*, the kangaroo was aesthetically domesticated, and the vogue of wonder associated with it was eroded. George Stubbs's painting and subsequent engraving of *The Kongouro from New Holland* (1773) was widely imitated and reissued. Christopher Lennox-Boyd identifies 111 distinct versions of the print issued between 1773 and the 1830s,[109] including an earthenware mug made by a Staffordshire pottery with a die-transfer print of the engraving.[110] Another appropriation of the image was for trade tokens (copper coins issued for local circulation in place of legal coinage) issued by managers of private menageries. Copper kangaroo tokens, with an image derived from Stubbs, were issued by George Pidcock, proprietor of the Royal Menagerie, and Thomas Hall, a taxidermist who operated a collection of mounted natural history specimens in City Road, Finsbury.[111]

The literal and figurative domestication of the kangaroo allowed it to be used in metropolitan political and cultural contexts, floating free of its location in natural history and spectacle. Two examples, both drawn from the field of literature, demonstrate this transition. The kangaroo was made the subject of an Aesopian verse fable by the Tory hack Richard Griffith called 'Tiger and Kanguroo' in 1792, who redeployed the animal in the debate on slavery. Griffith's kangaroo—"Belov'd by those who saw and knew/The properties of Kangaroo" –was evidently encountered in a public menagerie in London, exhibited alongside other exotic creatures, including a tiger.

> *A Kangaroo the other day*
> *(Who lately cross'd the wat'ry way*
> *To Britain's hospitable shore,*
> *From climes which Cook had sail'd t'explore)*

In manners gentle, temper kind,
Quick as a deer to catch the wind,
Beneath the roof where tigers fed,
Was doom'd to lay its faunish head.[112]

The "mild repose" of the gentle kangaroo is disturbed by the "surly" "roaring" of the tiger. In their ensuing dialogue, the kangaroo and the tiger both complain of their incarceration in the menagerie. The kangaroo, however, urges the tiger to accept his fate, and to enjoy the comforts of the menagerie, especially the provision of free and copious food. The tiger replies with a curse on these "timid counsels", preserving his right to fight for his liberty and his own victuals. He is shackled, he observes, because he is feared: set loose he would "reign the terror of the wood". The poem does respond to the experience of visiting the menagerie and encountering a kangaroo, but its primary intellectual engagement is with the contemporary debate on the slave trade. The kangaroo is made the mouthpiece, however ironic, of pro-slavery apologetics, defending the ease and comfort of the slave against the seditious agitation of abolitionists.

By the mid-1790s, the kangaroo was established as a recognizable emblem of New South Wales. That which made it distinctive and different, made it somehow typical of the convict settlement: it was as odd and anomalous an animal as the penal colony was a society. The radical poet Robert Southey, in his *Botany-Bay Eclogues* published anonymously in the daily newspaper *The Morning Chronicle* in September 1794, condemned the convict settlement, "ye savage lands! ye distant climes!" for their inherent cruelty and illegitimacy.[113] In poetic form, Southey's eclogues offered a critical revision of the conventional pastoral poem depicting a shepherd singing to his flock. In Southey's revision, the shepherd is conceived as an alienated rural laborer, reimagined and brought up to date by being depicted as a convict laborer on the state farm in Botany Bay. In the first eclogue, titled "Elinor", he narrates the woes of a clergyman's daughter seduced, abandoned, and coerced into prostitution ("the hireling prey/Of brutal appetite" ll. 48–49)—from which state has proceeded criminal dishonesty and then punishment, that of transportation ('A forlorn object, friendless, and unbewail'd' [l. 24]).

Welcome, ye wild plains,
Unbroken by the plough!—undelv'd by hand
Of patient rustic; where, for lowing herds,
And, for the music of the bleating flocks,
Alone is heard the Kanguroo's sad note
Deep'ning in distance. (ll. 69–74)

The kangaroo's alarm-call cough is the most distinctively alien contextual element in the poem: the sole item of Botany Bay's natural history famous enough to conjure an image in the urban reader.

The metropolitan construction of the kangaroo as both anomalous and yet typical of Botany Bay also made the genus a fitting and fertile emblem in the colony's self-representations. Barron Field's *First Fruits of Australian Poetry* (1819), the first volume of poetry published in the colony, featured a poem, "The Kangaroo", extolling the species as the "spirit of Australia". Like the early naturalists, Field found that in the kangaroo "Contradiction may be involv'd", describing the genus as a mixture of the squirrel and the hart, a mixture recalling mythical creatures like the mermaid, centaur and hippogriff.

For howsoe'er anomalous,
Thou yet art not incongruous,
Repugnant or preposterous.
Better proportion's animal,
More graceful or ethereal,
Was never follow'd by the hound,
With fifty steps to thy one bound.
Thou canst not be amended: no;
Be as thou art; thou best art so.[114]

While investing the animal with poetic and mythic nobility, Field also reinforces its status as a wonder of natural history. The poem delineates how the kangaroo is both an anomaly and yet also possessed of grace, beauty and virtue. In doing so, Field establishes a fertile analogy between the kangaroo and the colony itself, in this period beginning a process of establishing a civil society from the unprepossessing political materials of convict settlement. This mixture of wonder, beauty and anomaly continued to be used in representations of the kangaroo as an official emblem of Australia. The earliest example of Australian heraldic design, a commemorative silk pennant, known as the Bowman Flag of 1806, featured a kangaroo and an emu supporting the central blazon (in heraldic terms, the kangaroo was depicted 'regardant proper').[115] Bills of exchange issued by Union Bank of Australia in the 1850s were illustrated with a kangaroo (alongside a seated Britannia, a sailing ship and a wheat-sheaf).[116] When the Commonwealth Coat of Arms of Australia—the formal symbol of the nation –were granted by Edward VII in 1908, it depicted a heraldic shield of arms supported by a kangaroo (dexter) and an emu (sinister) standing on a grassy mound.[117]

In contrast to the national significance of the kangaroo's intangible heritage, the 'management' of the species is subject to continuing debate in Australia within science, public policy and popular culture. For most of the last century, kangaroos have been considered to be in competition with pastoral farming of sheep and cattle and have been subject to widespread control measures as 'noxious' pests. Although the major kangaroo species (*M. rufus, giganteus* and *fuliginosus*) are all assessed as lower risk ("least concern") on the International Union for the Conservation of Nature and

Natural Resources (IUCN) Red List of Threatened Species,[118] debate has continued as to the appropriate management of the species, both within national parks and nature reserves, and on pastoral rangelands, with management options ranging from conservation to commercial exploitation, through to culling and extermination. Shepherd and Caughley's "taxonomy of attitudes towards kangaroos," developed within a discussion of the management of the species, offers a cogent summary of the impact of the kangaroo's intangible natural heritage in the late 20th century.[119]

In its trajectory from capture to mounted and live display, this essay has followed the intangible natural heritage of the kangaroo in the later decades of the 18th century, showing how it was produced as an object of empirical knowledge in both metropolis and colony. The role of natural history, as a practice, a discourse and a discipline, was central to many of the key phases in this narrative. Natural history was a central legitimating discourse in the official documentation of Cook's voyage and the official print culture of the first fleet. It also played an important role in the museum and menagerie culture of the metropolis. It was clearly central to the question of nomenclature and classification played out in the official discourse of science and the Royal Society. Yet at each stage, the discourse of natural history contrasts and competes with non-scientific languages and practices: rhetorics of wonder and imagination, culinary preparations, field taxidermy, entrepreneurial display and heraldic emblemata. An intangible natural heritage of the kangaroo should recognize the diversity and sum of this complicated and contraindicated historical nexus of meaning.

NOTES

1. Peter Menkhorst and Frank Knight, *A Field Guide to the Mammals of Australia* (Melbourne: Oxford University Press, 2001). See also Terence Dawson, *Kangaroos: Biology of the Largest Marsupials* (Ithaca: Comstock Publishing Associates, 1995).
2. Laurajane Smith and Natsuko Akagawa (eds.), *Intangible Heritage* (London: Routledge, 2008).
3. Frank Hassard, "Intangible Heritage in the United Kingdom: The Dark Side of Enlightenment?" in *Intangible Heritage* (270–288).
4. John Berger, "Why Look at Animals?" (1977) in *About Looking* (London: Bloomsbury, 2009), 15.
5. John Hawkesworth, *An Account of the Voyages Undertaken by the Order of His Present Majesty for Making Discoveries in the Southern Hemisphere*, 3 vols. (London: William Strahan and Thomas Cadell, 1773), 3:577–578. This encounter, and the epistemological bewilderment it aroused, is the subject of Markman Ellis, "Tales of Wonder: Constructions of the Kangaroo in Late-Eighteenth-Century Scientific Discourse," in *Science and Exploration: European exploration of the South Pacific*, ed. Margarette Lincoln (London: Boydell and Brewer in association with the National Maritime Museum, 1998), 163–182.
6. Judy Egerton, "No. 143. The Kongouro from New Holland", in *George Stubbs, painter: catalogue raisonné* (New Haven, CT and London: Yale

University Press for the Paul Mellon Centre for Studies in British Art, 2007), 336–337. James Athenian Stuart, "The Kongaroo, an animal in New Holland—undescribed by any naturalist", exhibited at the Society of Artists, 1773. Susan Weber Soros (ed.), *James 'Athenian' Stuart: The Rediscovery of Antiquity* (New Haven; London: Yale University Press, 2006), 180.

7. *General Evening Post*, March 26, 1789; *Morning Chronicle and London Advertiser*, March 26, 1789; *Morning Post and Daily Advertiser*, March 26, 1789; *Felix Farley's Bristol Journal*, March 28, 1789.
8. The two ships had lost contact with each other soon after leaving Port Jackson, sailed via Cape Horn and made contact again in Rio de Janeiro. They sailed from Rio together and arrived in England separately, within two days of each other. News of their presence in Brazil, sent via Lisbon, was reported in newspapers on Monday, March 23, (*Public Advertiser*, March 23, 1789).
9. *The World*, March 26, 1789; *Stuart's Star*, March 27, 1789. The Borrowdale Transport, captain Hobsdon Reed, arrived at Plymouth on March 25, 1789 (*The Diary: or Woodfall's Register*, March 30, 1789): 'official dispatches from Commodore Phillip' were forwarded to the Admiralty as soon as possible (*General Evening Post*, March 31, 1789; *London Chronicle*, March 31, 1789).
10. *London Chronicle*, March 26, 1789; *Whitehall Evening Post*, March 26, 1789; *Stuart's Star*, March 26, 1789; *Morning Chronicle and London Advertiser*, March 27, 1789.
11. "Botany-Bay Expedition. Accurate Particulars", *General Evening Post*, Tuesday, March 31, 1789; "Farther Particulars of the Botany Bay Expedition", *London Chronicle*, March 28, 1789; "Botany-Bay Expedition. Accurate Particulars", *Whitehall Evening Post*, March 28, 1789; *Stuart's Star*, March 27, 1789; *Gazetteer and New Daily Advertiser*, March 28, 1789; "Accurate Particulars of the Botany Bay Expedition", *Morning Chronicle and London Advertiser*, March 28, 1789; *Bath Chronicle*, April 2, 1789.
12. "Account of the Expedition to Botany Bay", *Town and Country Magazine*, March 1789.
13. *Stuart's Star and Daily Advertiser*, March 30, 1789.
14. *The Diary or Woodfall's Register*, March 31, 1789.
15. Sydney Parkinson, *A Journal of a Voyage to the South Seas in His Majesty's Ship the Endeavour*, (London: Charles Dilly and James Phillips, 1784), 149.
16. John Hawkesworth, *An Account of the Voyages Undertaken by the Order of His Present Majesty for Making Discoveries in the Southern Hemisphere*, 3 vols. (London: William Strahan and Thomas Cadell, 1773), 3:577–578.
17. Raphael Cilento, "Sir Joseph Banks, F. R. S., and the Naming of the Kangaroo", *Notes and Records of the Royal Society of London* 26, no. 1 (1971): 157–161. Cilento refers to the Guugu-Yimidhirr as Gogo-Yimidir. See also Terence C. S. Morrison-Scott and Frederick C. Sawyer, "The identity of Captain Cook's kangaroo", *Bulletin of the British Museum (Natural History) Zoology* 1, no. 3 (1950): 43–50.
18. "Eora: Mapping Aboriginal Sydney 1770–1850", State Library of New South Wales (2006) online pdf, accessed July 20, 2009, <www.sl.nsw.gov.au/events/exhibitions/2006/eora/docs/eora-guide.pdf >.
19. John Hunter, *An Historical Journal of the Transactions at Port Jackson and Norfolk Island, with the Discoveries Which Have Been Made in New South Wales [. . .] Since the Publication of Phillip's Voyage* (London: printed for John Stockdale, 1793), 66. See also Mr Collins's vocabulary of the local language, collected by Collins and Phillip, much augmented by Capt Hunter (406–411): "Ban-ga-ray, *The red kanguroo*", (407); "Pa-ta-ga-rang, *The large grey kanguroo*" (410).

20. "Arthur Phillip to William Wyndham, Lord Grenville, Sydney, December 14 1791", in *Historical Records of New South Wales. Vol I, Part 2—Phillip. 1783–1792*, ed. Frank Murcot Bladen (Sydney: Charles Potter, 1892), 566.
21. Nancy Irvine, (ed.), *The Sirius Letters: The Complete Letters of Newton Fowell, Midshipman and Lieutenant aboard the Sirius, Flagship of the First Fleet on its Voyage to New South Wales* (Sydney: The Fairfax Library, 1988).
22. John Cawte Beaglehole (ed.), *The Endeavour Journal of Joseph Banks 1768–1771*, 2 vols. (Sydney: Angus and Robertson Limited, 1962), 2:94n.
23. William Bradley, *A Voyage to New South Wales. The Journal of Lieutenant William Bradley, R. N., of H. M. S. Sirius, 1786–1792. Reproduced in Facsimile from the Original Manuscript. With a Portfolio of Charts* (Sydney: Trustees of the Public Library of New South Wales in association with Ure Smith, 1969), 81. George Worgan's journal records the date as Saturday, February 9, 1788, in *Journal of a First Fleet Surgeon* (Sydney: The Library Council of New South Wales in association with the Library of Australian History, 1978).
24. Joseph Banks, *The Endeavour Journal of Joseph Banks 1768–1771*, ed. John Cawte Beaglehole, 2 vols. (Sydney: Angus and Robertson Limited, 1962), 2:94.
25. "John White to Joseph Banks, 18 November 1788", in *Sir Joseph Banks Papers: Brabourne Collection, 1786–1800*, vol. 3 (Sydney: Mitchell Library [State Library of New South Wales], Mss A78–2).
26. John Hunter, *An Historical Journal of the Transactions at Port Jackson and Norfolk Island, with the Discoveries Which Have Been Made in New South Wales [. . .] Since the Publication of Phillip's Voyage* (London: printed for John Stockdale, 1793), 66.
27. Arthur Bowes Smyth, *The Journal of Arthur Bowes Smyth: Surgeon, Lady Penrhyn 1787–1789* (Sydney: Australian Documents Library, 1979), 79. This specimen, preserved in alcohol (spirits) was returned to London with Lady Penrhyn, which left Port Jackson on May 5, 1788, and arrived in England, via Canton, in August 1789.
28. Matt Cartmill, 'Hunting and humanity in Western thought,' *Social Research* 62, no. 3 (1995): 773–786.
29. "Richard Johnson, A.L.S. to Joseph Banks, Sydney Cove, Port Jackson, New South Wales, 8 July 1788", (Sydney: Mitchell Library, Aj 1/4).
30. "Botany-Bay Expedition. Accurate Particulars, (Supplementary to Our Preceding Accounts.)", *The Morning Herald*, Saturday, March 28, 1789; *Whitehall Evening Post*, Tuesday, March 31, 1789; *The Public Advertiser*, Monday, March 30, 1789; "Farther Particulars of the Botany Bay Expedition", *London Chronicle*, Tuesday, April 7, 1789.
31. "Arthur Phillip, ALS to Joseph Banks, Sydney Cove, July 10 1788", (Sydney: Mitchell Library, Mss. ZC213). Harold B. Carter, *Sir Joseph Banks, 1743–1820* (London: British Museum [Natural History], 1988), 234.
32. "Arthur Phillip, ALS to Joseph Banks, Sydney Cove, July 2, 1788", (Sydney: Mitchell Library, Mss. ZC213).
33. "Denis Considen to Anthony Hamilton, Port Jackson, Nov 10 1788" [copy written in the hand of Banks], in *Sir Joseph Banks' Papers: Brabourne Collection 1786–1800*, vol. 3 (Sydney: Mitchell Library, Mss A78–2). See also Harold B. Carter, *Sir Joseph Banks, 1743–1820 (London:* British Museum [Natural History], 1988).
34. "Arthur Phillip, ALS to Lord Sydney, Sydney Cove, July 6, 1788", (Sidney: Mitchell Library, Dixson Mss Q162). See also "Arthur Phillip to Lord Sydney, Sydney Cove, July, 10 1788", in *Historical Records of New South*

Wales. Vol I, Part 2 –Phillip. 1783–1792, ed. Frank Murcot Bladen (Sydney: Charles Potter, 1892), 180: Phillip observed here that "This extraordinary animal makes the same use of his fore feet as the Monkey does".
35. *The World*, Saturday, March 28, 1789.
36. *The General Evening Post*, Saturday, April 4, 1789.
37. *The Diary: or Woodfall's Register*, Monday, March 30, 1789.
38. Advertised as this day published, *Whitehall Evening Post,* Thursday April 2, 1789. Tench's letter is reprinted in Tench, ed. Fitzhardinge, 333–336, with no mention of natural history.
39. *The Morning Post and Daily Advertiser*, Saturday, April 4, 1789. The published title was *The Voyage of Governor Phillip to Botany Bay*, and the imprint was dated 1790.
40. *London Chronicle,* Thursday, April 9, 1789.
41. British Library: pressmark C.47.110.
42. *Stuart's Star,* March 31, 1789.
43. "Arthur Phillip, ALS to Joseph Banks, Sydney Cove, September 26 1788", (Sydney: Mitchell Library, Mss. ZC213).
44. *The Morning Post and Daily Advertiser*, April 28, 1789.
45. "Panton Street and Oxendon Street", *Survey of London: Volume 20: St Martin-in-the-Fields, Pt III: Trafalgar Square & Neighbourhood* (1940), 101–103, accessed June 27, 2009, <http://www.british-history.ac.uk/report.aspx?compid=68420>.
46. *The Morning Post and Daily Advertiser*, Wednesday, April 29, 1789.
47. British Library, Lyson's Collectanea, C.103.k.11 [vol IV, 1, f. 22*].
48. John Hawkesworth, *An Account of the Voyages Undertaken by the Order of His Present Majesty for Making Discoveries in the Southern Hemisphere*, 3 vols. (London: William Strahan and Thomas Cadell, 1773), 3:577–578.
49. Arthur MacGregor, *Curiosity and Enlightenment: Collectors and Collections from the Sixteenth to the Nineteenth Century* (New Haven, CT. and London: Yale University Press, 2007), 143–149.
50. "Arthur Phillip, ALS to Joseph Banks, Sydney Cove, July 10 1788", (Sydney: Mitchell Library, Mss. ZC213).
51. The first record of the word in the *Oxford English Dictionary* (OED) is in *Taxidermy: or the Art of Collecting, Preparing, and Mounting Objects of Natural History. For the Use of Museums and Travellers* (London, 1820).
52. MacGregor, *Curiosity and Enlightenment*, 143–149.
53. Judy Egerton, "No. 143. The Kongouro from New Holland", *George Stubbs, Painter: Catalogue Raisonné* (New Haven and London: Yale University Press for the Paul Mellon Centre for Studies in British Art, 2007), 336–337.
54. *The World*, Tuesday, August 4, 1789.
55. MacGregor, *Curiosity and Enlightenment,* 146.
56. *A Companion to the Museum, (Late Sir Ashton Lever's) Removed to Albion Street, the Surry End of Black Friars Bridge* (London: [s.n.], Printed In The Year M.DCC.XC. [1790]). The kangaroo is not listed in this catalogue, perhaps because it had arrived so recently. The catalogue refers to other specimens located in the "Quadruped Room" (36), not apparently listed in the catalogue.
57. Thomas Pennant, *History of Quadrupeds*, 2nd ed., 2 vols. (London: B. White, 1781), 2:308.
58. "Banks to Piere Marie Auguste Broussonet, Soho Square, 1 May [1789]", *(Vol 3: The Middle Period, 1785–1799, Letters 1785–1790)*, in *Scientific Correspondence of Sir Joseph Banks, 1765–1820*, ed. Neil Chambers, 6 vols. (London: Pickering and Chatto, 2006), Letter 918, 3:486.

59. Harold Burnell Carter, *Sir Joseph Banks, 1743–1820* (London: British Museum [Natural History], 1988), 250.
60. *Edinburgh Evening Courant*, January 2, 1790; quoted in Michael Taylor, "An entertainment for the enlightened: Alexander Weir's Edinburgh Museum of Natural Curiosities", *Archives of Natural History* 19, no. 2 (1992), 153–167.
61. "Arthur Phillip, ALS to Joseph Banks, Sydney Cove, July 2 1788", (Sydney: Mitchell Library, Mss. ZC213).
62. "Arthur Phillip, ALS to Joseph Banks, Sydney Cove, Sept 26 1788", (Sydney, Mitchell Library, Mss. ZC213).
63. William Bradley, *A Voyage to New South Wales. The Journal of Lieutenant William Bradley, R.N., of H.M.S. Sirius, 1786–1792. Reproduced in Facsimile from the Original Manuscript. With a Portfolio of Charts* (Sydney: Trustees of the Public Library of New South Wales in association with Ure Smith, 1969), 117. On the Alexander's journey from Sydney to Batavia over half the crew died of scurvy.
64. *Lloyd's List*, September 6, 1791.
65. *Evening Mail*, October 24, 1791.
66. *Lloyd's List,* October 8, 1791.
67. *Evening Mail*, October 24, 1791.
68. Charles Bateson, *The Convict Ships, 1787–1868,* 2nd ed. (Sydney and London: Alfred H. and Alexander W. Reed, 1974).
69. *The Morning Chronicle*, December 6, 1791.
70. *Daily Universal Register*, January 9, 1787; Simon During, *Modern Enchantments: The Cultural Power of Secular Magic* (Cambridge, MA: Harvard University Press, 2002).
71. *Morning Chronicle*, January 12, 1792.
72. Handbill, 'The Wonderful Kangaroo,' *Lyson's Collectanea*, IV, 1, 22; also printed in *The Times*, November 16, 1791.
73. "Arthur Phillip to William Wyndham, Lord Grenville, Sydney, November 22 1791", in *Historical Records of New South Wales. Vol I, Part 2 –Phillip. 1783–1792*, ed. Frank Murcot Bladen (Sydney: Charles Potter, 1892), 559.
74. "Arthur Phillip to William Wyndham, Lord Grenville, Sydney, December 14, 1791", in *Historical Records of New South Wales. Vol I, Part 2 –Phillip. 1783–1792*, 566.
75. Mary Ann Parker, *A Voyage Round the World, in the Gorgon Man of War: Captain John Parker. Performed and Written by His Widow; for the Advantage of a Numerous Family* (London: John Nichols, 1795), 106.
76. *Star*, April 25, 1792.
77. "John Parker, ALS to Sir Joseph Banks, dated 'Gorgon Long Reach June 27th 1792,'" in *Banks Papers, Botanical and Horticultural (1789–1796)*, vol. 18 (Sydney: Mitchell Library), 85.
78. *Lloyd's Evening Post*, May 31, 1793; *The Observer*, Sunday, July 31, 1793.
79. *Gazetteer and New Daily Advertiser*, July 12, 1793. See also Richard, Drayton, *Nature's Government: Science, Imperial Britain, and the "Improvement" of the World* (New Haven and London: Yale University Press, 2000).
80. *The Observer*, Sunday, July 31, 1793.
81. *Lloyd's Evening Post*, Friday, May 25, 1792.
82. *Public Advertiser*, December 12, 1792.
83. *Public Advertiser*, December 12, 1792.
84. *Morning Chronicle*, January 22, 1793.
85. *The World*, February 5, 1793; September 24, 1793; *Morning Chronicle* November 28, 1793.
86. *The Sun*, December 6, 1793.

87. William Henry Brooke, "The anti-royal menagerie", *Satirist* (December 1, 1812), (British Museum Department of Prints and Drawings: Satires British, 1812, Unmounted Roy).
88. Carla Yanni, *Nature's Museums: Victorian Science and the Architecture of Display* (London: The Athlone Press, 1999), 14–36.
89. John Berger, "Why Look at Animals?' (1977), in *About Looking* (London: Bloomsbury, 2009), 21.
90. Everard Home, "Some observations on the mode of generation of the kanguroo: with a particular description of the organs themselves", *Philosophical Transactions* (1795):85 [221]–238.
91. Peter Jupp, *Lord Grenville: 1759–1834* (Oxford: Clarendon, 1985).
92. "Giovanni Valentino Mattia Fabbroni, Soho Square, 1 July 1796", (Joseph Banks, *Vol 4: The Middle Period, 1785–1799, Letters 1790–1799)*, in *Scientific Correspondence of Sir Joseph Banks, 1765–1820*, ed. Neil Chambers, 6 vols. (London: Pickering and Chatto, 2006), Letter 1362, 4:427–428.
93. *The Examiner,* October 11, 1828.
94. William Bingley, *Animal Biography; or, Anecdotes of the Lives, Manners, and Economy, of the Animal Creation,Arranged according to the System of Linnæus*, 3 vols. (London: printed for R. Phillips and sold by T. Hurst; E. Balfour, Edinburgh; and J. Archer, Dublin, 1803), 1:376.
95. Marcus Dunk, "Meet the lawnmower on legs: why wallabies are taking over England's gardens in leaps and bounds", *Daily Mail*, May 15, 2009.
96. William Beckford, *Azemia, a Novel* (London, 1798), 106.
97. Peter J. P. Whitehead notes "It is most curious that no early description was published based on the actual material", in "Zoological Specimens from Captain Cook's Voyages", *Journal of the Society for the Bibliography of Natural History* 5, no. 3 (1969): 161–201.
98. *Ibid*. See also Ronald M. Younger, *Kangaroo: Images through the Ages* (Sydney: Hutchinson Australia, n.d., [1980s]), 52, 59–65, 72. Solander's description was used by Johann Reinhold Forster in a letter to Linnaeus on July 26, 1776, with notes on "Yerbua Banksii Gigantea Novae Hollandiae dicta incolis *Kanguru* sic me olim descripta, & ni fallor Descriptionem tibi misi ante". Linnaean Correspondence in the Linnean Society Library, London, quoted in Whitehead, "Zoological Specimens", 195n.
99. Caroli Linnæi [Carl Linné], *Philosophica Botanica in qua explicatur fundamenta botanica* (Stockholm: Godofr Kiesewetter, 1751); quoted and trans. in Frans Stafleu, *Linnaeus and the Linnaeans:Tthe Spreading of Their Ideas in Systematic Botany, 1735–1789* (Utrecht: A. Oosthoek's Uitgeversmaatschappij for the International Association for Plant Taxonomy, 1971), 80.
100. Arthur Phillip, *The Voyage of Governor Phillip to Botany Bay*, 2nd ed. (London: John Stockdale, 1790), 129.
101. Thomas Pennant, *History of Quadrupeds*, 2nd ed., 2 vols. (London: B. White, 1781), 2:306–308.
102. There has been much discussion among historians and zoologists to clarify the identity of these species, a question made more difficult by the dispersal and destruction of the specimens. The last extant *Endeavour* specimen, a kangaroo skull, was preserved in the Museum of the Royal College of Surgeons but was destroyed by enemy action in World War II, although it is survived by a painting by Nathaniel Dance and a photograph made by the museum. Terence C. S. Morrison-Scott and Frederick C. Sawyer "The identity of Captain Cook's kangaroo", *Bulletin of the British Museum (Natural History) Zoology* 1, no. 3 (1950): 43–50, survey the research up to 1950 and conclude that the skull was that of a Great Grey Kangaroo *Macropus giganteus*, which they designate the photo-lectotype of *Macropus canguru* (Müller, 1776). The Latin

name is derived, by convention in zoology, from the first published zoological description, in this case Müller's German-language revision of Linnaeus (Phillip Ludwig Statius Müller, *Des Ritters C. von Linné [. . .] vollstandiges natursystem nach der zwölften Lateinischen Ausgabe [. . .] Supplementsband*, (Nürnburg, 1776), 62), where the kangaroo was in fact called *Mus canguru*. For many years the nomenclature was derived from Zimmermann's Latin *Specimen* (Eberhard A. W. Zimmermann, *Specimen zoologiae geographicae, Quadrupedum domicilia et migrationes sistens, &c.*, [Leyden: Theodorum Haak et Socios, 1777], 526), where this kangaroo was identified as *Jerboa giganteus* (Zimmermann, 1777). See also Averil Lysaght, "Captain Cook's kangaroo", *New Scientist* 1 (1957): 17–19 and Denis John Carr, "The Identity of Captain Cook's Kangaroo" in Denis John Carr (ed.), *Sydney Parkinson: Artist of Cook's Endeavour Voyage*, (London and Canberra: British Museum [Natural History] and ANU Press, 1983), 242–249.

103. Arthur Phillip, *The Voyage of Governor Phillip to Botany Bay*. (London: John Stockdale, 1789).
104. George Shaw, *Naturalist's Miscellany: or Coloured Figures of Natural Objects; Drawn and Described Immediately from Nature. Vivarium naturæ, sive rerum Naturalium, variæ et vividæ icones ad ipsam naturam depictae et descriptæ* (London: printed for Nodder and Co., 1790–1813). "The Great Kangaroo [Macropus Giganteus]" issued June 1, 1790, date from Frederick P. Nodder's engraved plate. George Shaw, *Cimelia Physica. Figures of Rare and Curious Quadrupeds,Bbirds, &c., Together with Several of the Most Elegant Plants, Engraved and Coloured by J. F. Miller: with Descriptions by G. Shaw* (London: Thomas Bensley for Benjamin and John White, and John Sewell, 1796).
105. George Shaw, *Musei Leveriani Explicatio, Anglica et Latina*, vol. 2 (London: Jacobi Parkinson, 1796), 38–39.
106. Home, "The mode of generation of the kanguroo", [221]–238.
107. Averil Lysaght, "Captain Cook's kangaroo", *New Scientist* 1 (1957): 17–19.
108. Home, "Mode of generation of the kanguroo", [221]–238.
109. Christopher Lennox-Boyd, Rob Dixon and Tim Clayton, *George Stubbs: The Complete Engraved Works* (Warren Farm House, Culham, Abingdon, Oxford: Stipple Publishing distributed by Sotheby's Publications, 1989), Nos. 361–472.
110. Kangaroo Mug, Staffordshire, ca. 1792, Earthenware (Canberra, National Gallery of Australia).
111. Peter Lane, *Kangaroos on English Tokens: A History and Catalogue of Late 18th Century Copper Tokens Depicting Kangaroos* ([Toorak Gardens, South Australia]: Peter Lane, 1984).
112. Richard Griffith, "Tiger and Kanguroo. April, 1792", in *Fables in Verse: Or, Present Life under Different Forms* (London: John Murray, 1793), 16–21.
113. "Botany-Bay Eclogue", *The Morning Chronicle*, No. 778, Thursday, September 18, 1794. The four *Botany-Bay Eclogues* were published together Robert Southey, *Poems by Robert Southey* (Bristol: N. Biggs for Joseph Cottle; London: G. G. and J. Robinson, 1797).
114. Barron Field, "The Kangaroo", in *First Fruits of Australian Poetry* (Sydney: Printed for Private Distribution [by George Howe], 1829), 9.
115. *Bowman Flag, 1806* (Sydney: Mitchell Library, 1806, Call number: XR 1).
116. Christopher Lennox-Boyd, Rob Dixon and Tim Clayton, *George Stubbs: The Complete Engraved Works* (Warren Farm House, Culham, Abingdon: Stipple Publishing distributed by Sotheby's Publications, 1989), Nos. 462, 360.
117. "Royal Warrant", in *Australian Government: It's an Honour: Australia Celebrating Australians*, accessed August 6, 2009<http://www.itsanhonour.gov.au/coat-arms/royal_warrant.cfm>.

118. *IUCN Red List of Threatened Species*, International Union for Conservation of Nature and Natural Resources, accessed May 13, 2010, <http://www.iucnredlist.org/about>.
119. Graeme Caughley, Neil Shepherd and Jeff Short (eds.), *Kangaroos: Their Ecology and Management in the Sheep Rangelands of Australia*, a joint project of New South Wales National Parks and Wildlife Service and Commonwealth Scientific and Industrial Research Organisation (CSIRO) Division of Wildlife and Rangelands Research (Cambridge: Cambridge University Press, 1987), 188–219; and Terence J. Dawson and Adam J. Munn, "How Much Do Kangaroos of Differing Age and Size Eat Relative to Domestic Stock?: Implications for the Arid Rangelands", in *Animals of Arid Australia: Out on Their Own?* (Sydney: Royal Zoological Society of New South Wales, 2007), 96–101.

5 "Project INH"

A Case Study of the Role of Museums in the Interpretation of Intangible Natural Heritage

B. Venugopal

Museums are traditionally recognized as repositories of tangible heritage.[1] In recent years, however, the intangible aspects of heritage have been gaining prominence through the efforts of the United Nations Educational, Scientific and Cultural Organization (UNESCO)[2] and others. Within the museum community, this move has been facilitated by the International Council of Museums (ICOM), where efforts to achieve parity between tangible and intangible heritage have gained momentum during the last decade.[3] The ICOM has, for instance, redrafted its definition of museums to include both tangible and intangible aspects of heritage.[4]

Most considerations around intangible heritage have, however, been restricted to the domain of culture. This is exemplified by the title of the UNESCO instrument of intangible heritage: *Convention for the Safeguarding of the Intangible Cultural Heritage 2003*. Although this document identifies five key domains in which intangible cultural heritage is manifest (oral traditions and expressions, performing arts, social practices, rituals and festive events, knowledge and practices concerning nature and the universe and traditional craftsmanship), natural heritage is relegated to the position of one out of five.

Although additional examples of natural heritage might be embedded within the other four domains of UNESCO's instrument, nature is relatively neglected in most discussions of intangible heritage.

Deliberations within ICOM have shown wide concern regarding the lack of due recognition for intangible natural heritage (INH) in the definitions of intangible heritage.

> *[There are] many ICOM members who work in science and natural history museums dealing with the cultural dimension and knowledge base of the collections and programmes. Documentation on traditional knowledge about biodiversity is an invaluable intangible element of natural heritage. The National Museum of Natural History in New Delhi and the Lawrence Durrell Zoo in Belize both integrate local knowledge and perceptions of science and the environment into their core activities.*[5]

In India, many museums and related institutions are actively involved in researching and interpreting intangible heritage. These include National Museum of Natural History, Indira Gandhi Rashtriya Manav Sangrahalaya (National Museum of Mankind), Indira Gandhi National Centre of Arts, Anthropological Survey of India and Council of Scientific and Industrial Research. A dominant focus on culture in India makes it difficult for nature to be given the place it deserves in discussions on intangible heritage. India has, however, been making rapid strides in traditional knowledge and biological diversity. These include the People's Biodiversity Registers, establishment of Biodiversity Boards and Traditional Knowledge Digital Library. However, these topics have been covered only sparingly in discussions on heritage in within the country.[6]

A common criticism of intangible natural heritage by Indian experts in intangible cultural heritage is the lack of good case studies from the natural sphere.[7] However, a search for case studies from natural heritage reveals important works on documentation of traditional knowledge about the medicinal properties of plants of India by early European colonizers. This chapter covers a work considered to be the earliest systematic and scientific documentation of intangible natural heritage in Asia.

PROJECT INH

The National Museum of Natural History (NMNH) took an early lead in intangible natural heritage by undertaking "Project INH". The main objective of this initiative was to showcase documentation of traditional knowledge about the indigenous medicinal practices of India by the early colonizers so as to make Indian policy makers aware of the importance of intangible natural heritage. The ultimate goal was to achieve equity between natural and cultural topics in discussions of intangible heritage. The main focus of Project INH was Van Rheede's seminal work *Hortus Malabaricus,* the earliest systematic documentation of intangible natural heritage in Asia. Project INH included two national meetings on intangible natural heritage and *Hortus Malabaricus,* respectively, as well as the establishment of a Museum Gallery "*Early Plant Heritage of India*".

Calicut Conference 2007 (Appendix 3): A national conference on "Intangible Natural Heritage and Museums" was held in Calicut between April 18 and 20, 2007. About 100 experts from various museums and related organizations in India participated, as well as representatives from the Netherlands (National Herbarium) and UNESCO. A Calicut Charter on Intangible Heritage and Museums was released at the end of the conference (Appendix 3). It highlighted the importance of intangible natural heritage and the significance of *Hortus Malabaricus* and suggested the recommendation "to consider the [17th century] publication *Hortus Malabaricus* . . . as the earliest Asian example of systematic, scientific documentation of intangible natural heritage through the method of oral transmission of traditional knowledge about the plant medicinal use prevalent at that time in Malabar, belonging to the present Kerala State".

Kochi Workshop 2008: As a follow-up to the Calicut Conference, a national workshop on *Hortus Malabaricus* was held between March 14 and 16, 2008 in Kochi, a city where Hendrik van Rheede, the author of the work, undertook his preparations. One of the keynote speakers at the conference was Dr Annamma Spudich, who had curated an exhibition on *Such Treasure and Rich Merchandise* at the National Centre for Biological Sciences (NCBS) Bangalore (2008), which included many exhibits on *Hortus Malabaricus*.

Museum gallery 2009: An exhibition curated by Stanford University's Dr Annamma Spudich, which was on show at the NCBS for a few months was re-structured as a new Gallery on "*Early Plant Heritage of India*" and opened to public at the Regional Museum of Natural History (RMNH) Mysore in January 2009. The gallery highlighted the importance of *Hortus Malabaricus*.

Even though the first volume of *Hortus Malabaricus* was published in 1678, few people are aware of the existence of the publication or its importance. Dr Kattungal S. Manilal, professor of botany at the University of Calicut, India, has worked for about 40 years studying it and then translated it into English and Malayalam. This was published by the University of Kerala in 2003 and 2008, respectively.

Project INH involved efforts to highlight the importance of *Hortus Malabaricus* as a classic example of documentation of intangible natural heritage, applicable to the 2003 UNESCO convention. Advice and help from Dr Manilal, Dr Spudich and Dr G. V. Subrahmanyam (Ministry of Environment and Forests, Government of India) made Project INH a success. The project also led to the revival of interest in another publication *Jardin de Lorixa*, which was compiled about a century later, inspired by the success of *Hortus Malabaricus*. This is a 14-volume manuscript published in 1725 by L'Empereur and now housed in the library of *Musée National d'Histoire Naturelle* in Paris.

HORTUS MALABARICUS

Hortus Malabaricus is a unique work, in 12 volumes, published in the Netherlands between 1678 and 1693 (Figure 5.1). *Hortus Malabaricus* was one of the main sources of Linnaeus for his knowledge of the tropical flora of Asia. After this, Van Rheede's work was consulted by taxonomists in their critical studies of Linnaean species. The publication documents the traditional knowledge of the plants of Malabar (currently Kerala State in India), including multiple uses and medicinal properties. The work has been a major contributor to the development of modern botany. Hendrik

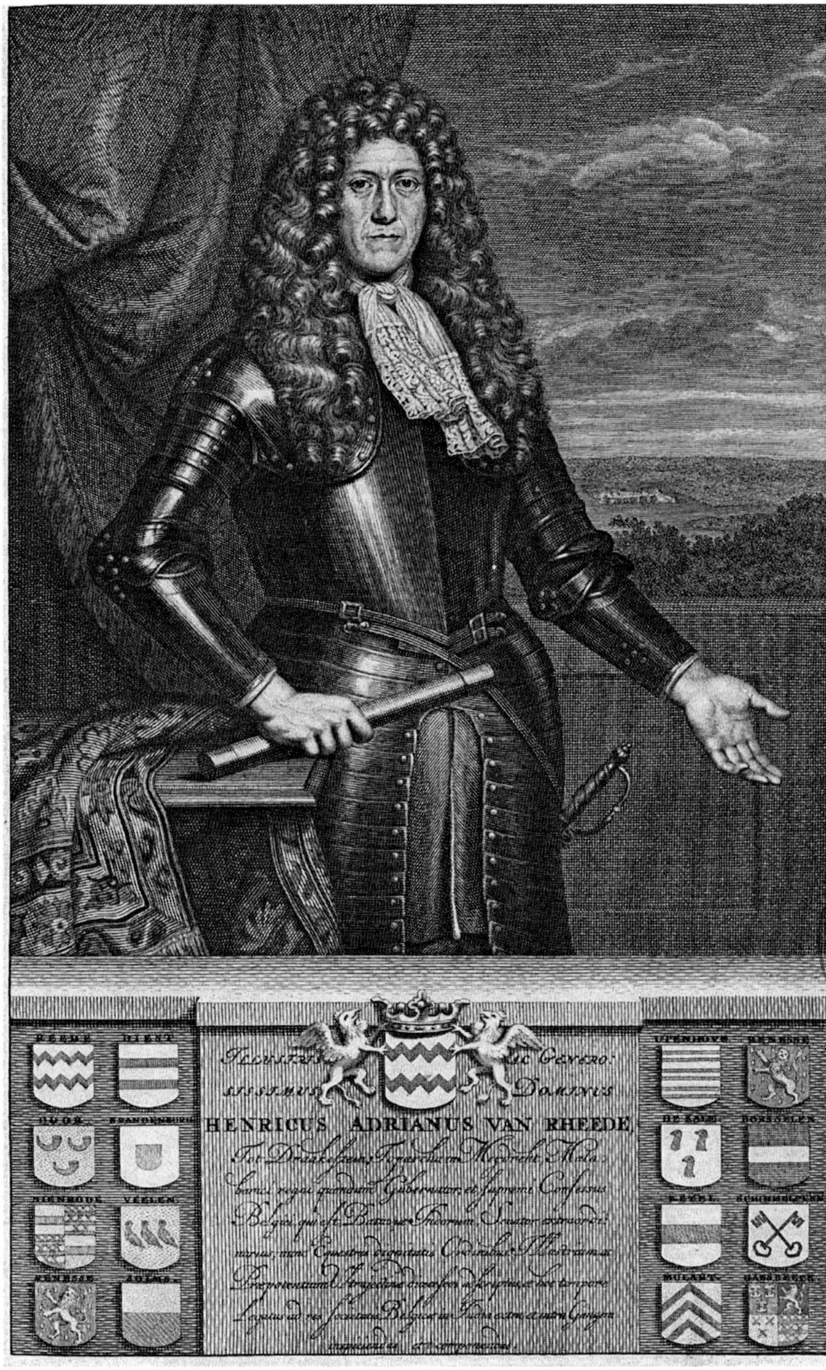

Figure 5.1 Frontispiece to volume 1 of *Hortus Malabaricus,* 1678.

Adriaan van Rheede tot Draakenstein (1636–1691) was the compiler and promoter of *Hortus Malabaricus*. Born of a noble family in Utrecht, Netherlands, he went to India as a soldier in the army of the Dutch East India Company. Van Rheede fought in Malabar Kingdom and was rapidly promoted, becoming the Dutch Governor of Cochin in 1663 and the Commander of Malabar in 1669.

Hortus Malabaricus is arguably the only book on botany to have influenced the political history of a country, its creation catalyzed by a combination of scientific curiosity, political and economic drivers.[8] The European colonizers' eagerness to make discoveries while exploring new countries, especially within Asia, in large part drove the desire to document the natural wealth of Malabar. Van Rheede's duties took him on long tours of inspection through the green countryside of Malabar, and he was fascinated by the diversity of plant life there. Van Rheede interrogated local Indians about the names of their flora and their uses, particularly medicinal.

In 1669 the Dutch-controlled island of Batavia, now Jakarta, made a request from the Government of Ceylon to investigate the availability of medicinal plants there, in an effort to reduce the necessity of importing drugs from the Netherlands. Medicines were expensive and often deteriorated during the long trip from Europe. Van Rheede was aware of this and hoped also to find similar work in Malabar, suspecting (rightly) that a survey of the Malabar flora would be a useful step towards self-sufficiency of the Dutch stationed there. Thus, from his initial curiosity emerged a desire to systematize and extend what had previously been a random acquisition of botanical knowledge for the benefit of the Dutch East India Company, as well as botanists back in Europe.[9]

However, at this time, a rivalry existed between Van Rheede and his superior, Admiral Van Goens over the most appropriate location for a second Dutch capital in Asia, after Batavia. Van Rheede was for selecting Colombo on the island of Ceylon, today Sri Lanka, and Van Goens for Cochin, in the Malabar Region of Southern India. Van Rheede's familiarity with the flora and society of Malabar led him to feel that its self-sufficiency in essential requirements like medicine, food, and timber for shipbuilding, was critical in the event of siege. The availability of a skilled local workforce also contributed to Van Rheede favoring Cochin over Colombo.[10] Thus, Van Rheede's decision to compile a book on the medicinally useful plants of Malabar was, at least in part, to demonstrate that Malabar was more suitable than Ceylon.

Rising to the competition, in 1678, Van Goens recruited the German physician and botanist Paul Hermann to write a flora of Ceylon to rival *Hortus Malabaricus*. Although Hermann was a preeminent scholar, the next year to become chair of botany at the University of Leiden, the work never appeared. After the publication of the first volumes of *Hortus Malabaricus*, the high government at The Hague decided in favor of making Cochin their Eastern Capital, supporting Van Rheede and disregarding

Van Goens. This was an extremely important move. As Kattungal Manilal points out, "Had the decision been otherwise, the subsequent history of [the entirety of] India would have been very different, with its own effect on the other European powers who followed the Portuguese to India".[11]

One of the book's major influences was on the religious sphere in Malabar, on both Hinduism and Christianity.[12] At this time, the Dutch were Protestants and the Portuguese their Catholic rivals. Hence, when the Dutch defeated the Portuguese in Malabar, most Catholics present there fled or went into hiding. Even though Van Rheede was a Protestant, he helped Carmelite Catholic Father Mathew to build its first church, thereby securing the clergyman's much-needed local botanical knowledge for the initial work of *Hortus Malabaricus.*

> *It was Van Rheede's scientific spirit of plant explorations in Malabar, which became the cause for the construction of the first Carmelite church in Malabar and India, at Chathiath near Cochin. Later, Fr. Mathew used the goodwill of Van Rheede to build a second Carmelite church also in the same year at Verapoly [Varapuzha], thus cementing the foundation of Carmelite Missionary work in Malabar.*[13]

Hinduism has a strongly entrenched caste system and, at this time, the local Malabar Hindu community observed strict caste formalities. The high caste Brahmins and Nairs, just below them in the hierarchy, enjoyed specific privileges. Others, considered lower castes, were treated as untouchables with no access to education or the ability to worship in temples. Even walking along public roads was forbidden, despite the fact that taxes from the lower castes contributed to maintaining them. Social humiliations included being prohibited from walking within 40 yards of a high caste person, saying nothing or touching or communicating with him.

"Ezhava" was one of the untouchable castes. Unconcerned about caste, Van Rheede enlisted the assistance of a physician of the Ezhava caste, Itty Achuden (now referred to as Itty Achuthan), [14] to ensure the authenticity of *Hortus Malabaricus*. This had implications far beyond the robustness of the work.

> *By considering the professional service of this untouchable physician more acceptable, authentic and superior to those of a Christian Missionary physician Fr. Mathew and the three Konkani Brahmins recommended by the ruling King, Van Rheede perhaps symbolized the alien concept of recognition of personal merit and the dignity of all men . . . This should justifiably be considered as one of the biggest milestones in the social history of medieval Malabar, with far reaching effects.*[15]

Hortus Malabaricus also represents a landmark in the history of Indian languages. Two languages prevalent in India, Malayalam and Konkani, appeared in print for the first time in *Hortus Malabaricus*. These appeared as signed certificates acknowledging the involvement of various individuals in the compilation of the book. The Malayalam language was written in kolezhuthu script by Itty Achuden, and the Konkani language was written in nagri script by three Konkani Brahmins (Ranga Bhat, Vinayaka Pandit and Appu Bhat). Another certificate in Malayalam (in Aryezhuthu script) by Emmanuel Carneiro, the official interpreter who knew Malayalam, also vouches for the participation of Itty Achuden in the compilation of the book. Thus, *Hortus Malabaricus* is also perhaps the only instance ever in which both the scripts of Malayalam appear together in the same publication (Figure 5.2). Carl Linnaeus accepted many Malayalam plant names given by Van Rheede as the generic of specific epithets in the new binomial system of his *Systema Naturæ* (1735), ensuring their lasting contribution.

Hortus Malabaricus was highly influential on the development of plant classification. In the 150 years before the preparation of the work, the European system of classification of plants was dominated by official government patronage, founded on teaching gardens in major universities and associated herbaria, known as *Horti Sicci* (dry gardens). This information

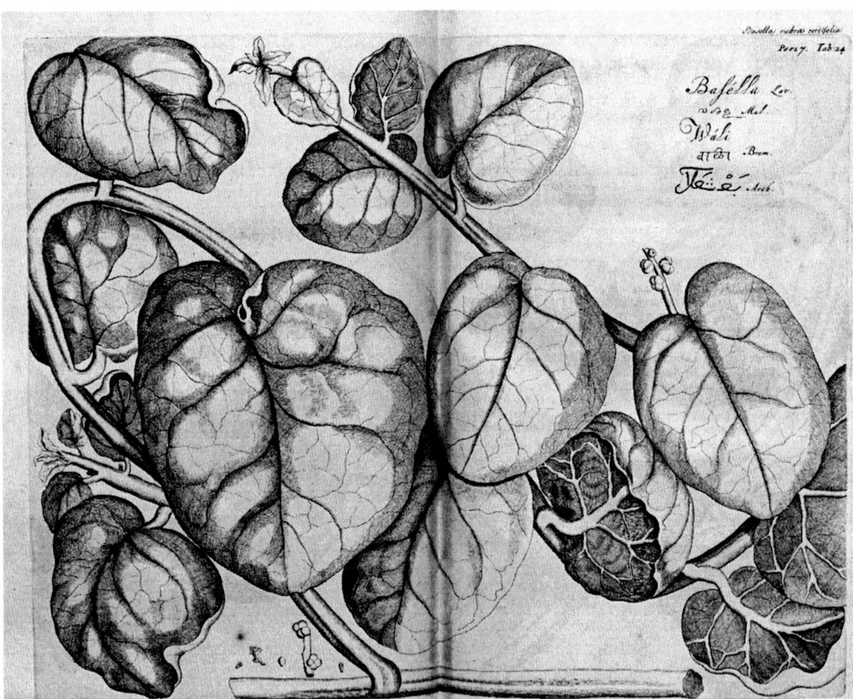

Figure 5.2 An illustration of *Basella rubra*, (=*alba*), or Malabar spinach, from volume 7, page 47, plate 24, of *Hortus Malabaricus*, 1688, showing the name in various languages, including Konkani and Malayalam.

was compiled into *materia medica*: extensive treatises on the therapeutic properties of substances used for healing. These were the earliest resources for the European system of plant classification.

During European colonization of India, travelers were encouraged to collect material to extend European *materia medica*. This led to the publication of the first major European book on Asian botany, *Coloquios dos simples e drogas cousas medicinas da India (1563)*, the work of Garcia da Orta, a Portuguese physician who lived in Goa. It was followed closely by Christobel d'Acosta's *Tracado de las drogas y medicine de las Indias Orientalis (1578)*. Although Garcia da Orta's *and* Christobel d'Acosta's works were already published by the time of creation of *Hortus Malabaricus*, each dealt only with about 100 species of plants with fragmentary descriptions of plants growing around Goa and Cochin, locations about 300 km apart. *Hortus Malabaricus* is set apart by the fact it was the first comprehensive regional floristic survey in Asia.[16] A strong synergy exited between the works and their timing, both for the acquisition of knowledge, and with respect to the development of overseas expansion.

> *The close association between Clusius and Garcia d'Orta and the connections between Van Rheede and Dutch botanical establishment ensured that the diffusion of botanical knowledge between south-west India and the Leiden botanic garden became central to the whole relationship between European and Asian constructions of nature. Because of this, two main texts can be said to lie at the core of the relationship between European colonial expansion and the diffusion of botanical knowledge.*[17]

For the sake of accuracy, Van Rheede, as had da Orta before, made extensive use of Ayurvedic and Ezhava medical and biological knowledge. This had the effect of promoting 'traditional' knowledge above the preexisting Arabic, Brahminical and European works, creating "profoundly indigenous texts . . . organised largely according to non-European precepts".[18] Using traditional knowledge, irrespective of caste, at a time when expanding Europe was hungry for innovation, is a major reason this book was able to have its extent, and breadth, of impact.

Compiled more than a century after da Orta wrote the *Coloquios*, Van Rheede went through the same process of rejecting Arabic classification and nomenclature, as well as European knowledge, in favor of a more rigorous adherence to local systems of classification. It took only two years for him to reject the methodologies of plant description represented by the *Viridarium Oreintale* of Father Mathew of St Joseph. Van Rheede's own writing shows clearly that Achuden played a role as ultimate arbiter of accuracy and correct identification. This was because the Dutchman soon found that the botanical learning of the Konkani Brahmins was, in fact, quite weak and entirely dependent on the restatement of dictums from old texts. For any useful field identification and collection of particular plants desired by Van Rheede and the *Hortus Malabaricus* 'board' (see below), the Brahmins were forced to rely on the much greater field knowledge of their low-caste servants. Their natural science knowledge was, as far as Van Rheede was concerned, merely academic. Thus, it made sense to bypass the Brahmins.

Van Rheede probably never made a herbarium of Malabar plants, instead relying on plant identification skills of the tree-climbing Ezhava collectors of the 'toddy-tapper' caste. Among this group were families of Vaidyar traditional doctors, whose occupation was passed down from father to son, along with the written information on medicinal uses of plants. Achuden and his fellow Ezhava tree-climbers selected the plants to be drawn and included, described their names for the plants and their uses. Thus, the contents of *Hortus Malabaricus* were strongly influenced by the choices of the Ezhava collectors. Most importantly, from the standpoint of tropical botany that was to flow on from this work, many insights of the Ezhava into species are reflected the structure of their names:

> *. . . which have the same stem and to which one or more prefixes are added: for example Onappu, Valli-onappu and Tsjeri-onappu. The names also provide considerable amount of incidental sociological material. In Onappu 'Onam' is the harvest festival in which this particular flower would be used. Thus, the names preserve the true social affinities of the plant name, instead of isolating them in a context less arbitrary category, as well as allowing, probably, a true affinity in terms of pharmacological properties.*[19]

These names fed directly into the plants' classification.

Many historically and botanically important texts have relied heavily on *Hortus Malabaricus*. "*Species Plantarum*" published in 1753 by Carl Linnaeus (1707–1778) described all the then known plant species of the world, presenting a codified system where all plant species then known were included with binomials, often with key characters, reference and information on their habitats. Linnaeus used the Ezhava classification to establish 240 entirely new species and, among the many old books used, *Hortus Malabaricus* by Van Rheede was one of only two that commanded his explicit faith in its veracity.[20]

Linnaeus used 258 of the Ezhava plant names in *Hortus Malabaricus* species binomials for 255 species in 149 genera in *Species Plantarum*. He later used another 95 Rheedean elements, for a total of about 350 species. Of these, 55 species in 49 genera were created solely based on Van Rheede's work. Linnaeus used many more to coin binomials either directly or after Latinizing them. As a result, Malayalam contributed more than any other Indian language to *Species Plantarum*.[21] Adanson followed Linnaeus' for his 1763 classification,[22] as did many others who followed.

The documentation methodology used preparing *Hortus Malabaricus* would not have succeeded as well in Europe of the day as it did in Asia. Van Rheede describes his methods in volume III of *Hortus Malabaricus*: two phases of compilation and documentation, the first phase involving European methodology, which did not succeed and was rejected, followed by a second, which focused on local traditional knowledge.

FIRST PHASE: EUROPEAN METHODOLOGY

Van Rheede initially enlisted the help of Father Mathew of St Joseph (an Italian Carmelite missionary, a physician from Naples claiming familiarity with medicinal plants) for the preparation of *Hortus Malabaricus*. However, Van Rheede soon began to have reservations about the work being turned by Father Mathew, as well as the utility of the unsophisticated illustrations drawn by pen and prepared in large part from memory. The slow progress also exasperated Van Rheede. Paul Hermann, who visited Cochin in 1674 on way to Ceylon, also dismissed Father Mathew's work as substandard.[23]

> *When I got stuck, there arrived in Malabar from the island of Ceylon the famous Paul Hermann, doctor of medicine, who had just been appointed Professor of Botany at the illustrious University of Leiden . . . with modest kindness he saw fit to give us to understand that in this way it would hardly become a book of any importance. . . . at least at his* [Father Mathew's] *advice I solicited for part of the work the aid of a highly revered man, Johannes Casearius, a vigilant minister of God's World in the church united in Christ in the city of Cochin.*[24]

Thus, Van Rheede's initial attempts to document the plant wealth of Malabar using the European methodology to which he was accustomed was destined for failure, and, if he were to accomplish his mission, he would have to look elsewhere.

SECOND PHASE: FOCUS ON LOCAL TRADITIONAL METHODOLOGY

Van Rheede then turned, pragmatically, to the principal source of botanic knowledge in the region: the people who themselves lived among, and used, the plants.

> *Some physicians, both Brahmans and other, at my orders made in their own language lists of the best known and frequently occurring plants, on the basis of which others again divided the plants according to the*

> *season in which they attracted notice, because they bore either leaves or flowers or fruit. This catalogue according to the time of the year was given to certain men who were experts in plants, who were entrusted with collecting for us finally from everywhere the plants with the leaves, flowers and fruit for which they even climbed the highest tops of trees. Having generally divided them into groups of trees, I sent them to some forest. Three or four painters, who stayed with me in a convenient place, at once accurately depicted the living plants readily brought by the collectors. To these pictures a description was added nearly always in my presence. Moreover, I had courteously asked all princes of Malabar kindly to inform us of plants which were rather well known by their form and their use . . . The plants which had thus been collected by botanists, depicted by painters, described by capable men, were subjected, if this was worthwhile, to the examination of skilful physicians and botanists, whom I had convoked for the purpose, both from my staff and from among people of some reputation. Naturally I took care that the pictures of plants were shown to this board, which sometimes consisted of fifteen to sixteen scholars and that by means of an interpreter they were asked whether they knew those plants and their names and curative virtues. The answers were recorded in the commentarium.*[25]

In his preface, Van Rheede mentions the role played by the local medical experts, referring to them simply as "some physicians" without specifying them. However, the volume I certificates mentioned previously give acknowledgement to the four local collaborators: Ezhava physician (Itty Achuthan) and the three (Konkani) Brahmin physicians (Appu Bhat, Ranga Bhat and Vinayaka Pandit).

Hortus Malabaricus is an ideal example of intangible natural heritage. Van Rheede notes that the medicinal uses of the plants described in it are quoted from the "family books" of the Collat Vaidyars (modern: Collat Vaidyan), supported by Itty Achuden's own practical experience. These family medical books have all been lost. The Konkani Brahmins were reported to have referred to their ancestors' books for describing the medicinal properties of plants. However, efforts to trace these ancestral books have proven futile. It is safe to say that if Van Rheede had not compiled the information as published in *Hortus Malabaricus*, the valuable documentation on the hereditary ethnomedical knowledge of Malabar would have been totally lost.

The traditional knowledge about the plant medicinal uses prevalent in the 17th century in Malabar are described in the book *Hortus Malabaricus* by employing various systematic and scientific methods of documentation including oral narrative. No such systematic records from Asia are available prior to this documentation. Thus, *Hortus Malabaricus* is considered as the earliest example of systematic and scientific documentation of intangible natural heritage from Asia, and provides a foundation for further work.

The knowledge held by the hereditary physicians of Malabar was exclusive to their families and recorded in palm leaf manuscripts held by them. Over the years most of these manuscripts have disappeared. The Hortus Malabaricus *is an invaluable record of these medical traditions as they existed in the 17th century. Recently Dr. D. Nicolson of the Smithsonian Institution, Dr. C. R. Suresh and Prof. K. S. Manilal of Calicut University updated and correctly identified the plants form the* Hortus Malabaricus *using modern botanical nomenclature to make the information accessible for further scientific study. Similar efforts to increase the accessibility of the knowledge recorded in the other books in this exhibit would provide valuable opportunities to study again these forgotten medical traditions.*[26]

APPENDIX 1

Shanghai Charter

We, the coalition of participants, at the 7th Asia Pacific Regional Assembly of the International Council of Museums convened between 20–24 October 2002 in Shanghai, affirming the significance of creativity, adaptability and the distinctiveness of peoples, places and communities as the framework in which the voices, values, traditions, languages, oral history, folk life and so on are recognized and promoted in all museological and heritage practices, recommend that museums as facilitators of constructive partnerships in the safeguarding of this heritage of humanity:

1. Affirm the rich cultural diversity of the Asia Pacific, including the concerns of race, ethnicity, color, gender, age, class, faith, language, sexual orientation and regional identities;
2. Establish interdisciplinary and cross sectoral approaches that bring together movable and immovable, tangible and intangible, natural and cultural heritage;
3. Address the challenges and threats posed by globalization and develop approaches to maximize on the opportunities provided by cultural, technological and economic globalization;
4. Develop documentation tools and standards in establishing holistic museum and heritage practices;
5. Initiate pilot projects that demonstrate methodologies for making inventories of intangible heritage resources through community participation;
6. Ensure efforts towards the conservation, presentation and interpretation of intangible heritage in an authentic manner that is consistent with the local character;

7. Develop public programs and visitor management strategies that comply with the laws, conventions and regulations dealing with the conservation of heritage resources of significance, and respect for the rules and protocols of the community groups as custodians of intangible heritage;
8. Encourage cross cultural understanding and meaningful exchanges for the promotion of peace and harmonious societies;
9. Make use of all media formats such as print, audio visuals, film and video, digital and telecommunication technologies;
10. Assess and address training needs and capacity building for integration of tangible and intangible heritage management;
11. Offer inclusive interpretation in appropriate languages where possible employing local custodians of intangible heritage resources;
12. Promote the active participation of public and private sectors to maximize the use of local expertise, resources and opportunities and diversify the resource base for the effective safeguarding of all heritage resources;
14. Establish criteria and methodologies for the integration of tangible and intangible heritage in museums and other heritage institutions;
15. Support UNESCO's efforts for the safeguarding and promotion of intangible heritage through its various programs and stress the importance of the inputs of professional bodies in the preparation of an international convention for safeguarding intangible cultural heritage.

Signed by Professor Zhang Wenbin, President, ICOM China, Dr. Jacques Perot, President ICOM and Professor Amareshwar Galla, President ICOM Asia Pacific.

APPENDIX 2

Seoul Declaration of ICOM on the Intangible Heritage

Resolution no. 1

Considering the undeniable importance of intangible heritage and its role in the preservation of cultural diversity, the 21st General Assembly of ICOM, held in Seoul on 8 October 2004

1. Endorses the 2003 UNESCO Convention on the Protection of Intangible Cultural Heritage;
2. Urges all governments to ratify this convention;
3. Encourages all countries, and especially developing countries where there is a strong oral tradition, to establish an Intangible Heritage Promotion Fund;
4. Invites all relevant museums involved in the collection, preservation and promotion of intangible heritage to give particular attention to

the conservation of all perishable records, notably electronic and documentary heritage resources;

5. Urges national and local authorities to adopt and effectively implement appropriate local laws and regulations for the protection of intangible heritage;
6. Recommends that museums give particular attention and resist any attempt to misuse intangible heritage resources and particularly their commercialization without benefits to the primary custodians;
7. Urges regional organizations, national committees and other ICOM bodies to work closely with local agencies in the development and the implementation of such legal instruments and in the necessary training of staff responsible for effective implementation;
8. Recommends that all training programmes for museum professionals stress the importance of intangible heritage and include the understanding of intangible heritage as a requirement for qualification;
9. Recommends that the Executive Council, working with the International Committee for the Training of Personnel (ICTOP), introduce the necessary adjustments as soon as possible into the ICOM Curricula Guidelines for Museum Professional Development (1971, latest revision 1999);
10. Decides that this Resolution shall henceforth be known as the "Seoul Declaration of ICOM on the Intangible Heritage".

APPENDIX 3

Calicut Charter on Intangible Heritage and Museums

The participants of the National Conference on Intangible Natural Heritage and Museums organized by the RMNH Mysore (Ministry of Environment and Forests, Government of India) in collaboration with the Ecotourism Directorate (Government of Kerala) during April 18–20, 2007, in Calicut (India), in which about 100 experts (including from UNESCO and the Netherlands) participated, recommend the following, which may be called as the "Calicut Charter on Intangible Heritage and Museums":

1. Support the concerns of the ICOM (International Council of Museums) that Natural heritage and Cultural heritage are two equally important component parts of Intangible Heritage,
2. Recommend to adopt the Indian Model on Intangible Heritage, which was originally presented in the ICOM-ASPAC workshop on "Intangible Heritage, Museums and Globalization", held in Shanghai, China in October 2002. This Indian model on Intangible Heritage is based on the Model on Access and Benefit Sharing involving the original stakeholders, the Kani tribes in India,

3. Recommend to consider the publication Hortus Malabaricus (compiled by Van Rheede the Dutch Governor of Malabar with headquarters in Cochin and published from Netherlands during 1678–1693) as the earliest Asian example of systematic, scientific documentation of intangible natural heritage through the method of oral transmission of traditional knowledge about the plant medicinal use prevalent at that time in Malabar, belonging to the present Kerala State,
4. Recommend that the People's Biodiversity Registers (PBR) already available in India be considered as examples of community involvement in the documentation of intangible natural heritage,
5. Recommend that the National Biodiversity Authority (Ministry of Environment and Forests, Government of India) establish a National Committee on Intangible Natural Heritage in collaboration with the NMNH,
6. Recommend that the various agencies of the Government of Kerala especially the Department of Forests establish a State Institute of Intangible Natural Heritage to coordinate all activities regarding the collection, conservation, documentation, research, exhibition and management of *Hortus Malabaricus,*
7. Appreciate the efforts of the Centre for Heritage Studies, Thripunithura to establish a Department of Intangible Heritage, a Heritage Garden (based on *Hortus Malabaricus*) and a Museum Gallery on *Hortus Malabaricus*/Van Rheede,
8. Recommend to the professional organizations in Museology such as MAI (Museums Association of India), ICOM (International Council of Museums), and INC-ICOM (Indian National Committee of ICOM) to give wide publicity to the Calicut Charter,
9. Recommend to the various Ministries of the Government of India (such as the Ministry of Forests & Environment, and the Ministry of Culture) to consider the Calicut Charter in all discussions on Intangible Heritage.

Signed on behalf of the participants of the Conference: Dr B. Venugopal, Coordinator of the National Conference.

NOTES

1. International Council of Museums, Statutes. Adopted by the 16th General Assembly of ICOM (The Hague, Netherlands, September 5, 1989). http://icom.museum/what-we-do/programmes/intangible-heritage.html (accessed 30 May 2011).
2. UNESCO has been involved in the normative processes related to intangible heritage. See, for instance, the Proclamation of Masterpieces of the Oral and Intangible Heritage of Humanity (2001, 2003 and 2005) and the Convention for the Safeguarding of the Intangible Cultural Heritage (2003). On October 17, 2003, following the Second Proclamation of Masterpieces of the Oral and

Intangible Heritage of Humanity, the General Conference of UNESCO adopted the Convention for the Safeguarding of the Intangible Cultural Heritage.

3. This topic has been discussed in detail in the ICOM meetings in Barcelona, Shanghai (Appendix 1) and Seoul (Appendix 2) http://www.chin.gc.ca/Applications_URL/icom/shanghai_charter.html (accessed 30 May 2011 & http://icom.museum/what-we-do/resources/publications-database/publication/2004-seoul-icom-icme-conference-museums-and-intangible-heritage-seoul-korea-2-01-october-2004.html (accessed May 30, 2011).
4. "A Museum is a non-profit, permanent institution in the service of society and its development, open to the public, which acquires, conserves, researches, communicates and exhibits the tangible and intangible heritage of humanity and its environment for the purposes of education, study and enjoyment". ICOM, Statutes. Adopted by the 22nd General Assembly of ICOM (Vienna, Austria, August 24, 2007).
5. Amareswar Galla, "Frequently asked questions about intangible heritage", *ICOM News* 4 (2003), 4.
6. The ICOM Seoul Meeting suggested having the first "Capacity Building Workshop on Documentation of Traditional Knowledge, Museums and Intangible Natural Heritage" in Mysore (India), based primarily on a paper on intangible natural heritage from India (Bargaviamma Venugopal, 2001) presented at the Asia Pacific Workshop on Museums and Intangible Heritage (Shanghai, China, 2001). Although this never materialized, Dr Amareshwar Galla organized an ICOM "International Workshop on Traditional Knowledge Systems, Museums and Intangible Natural Heritage in South Asia" in Hyderabad, India, during February 3–7, 2008.
7. Bargaviamma Venugopal, "Museums and Intangible Heritage: Recent Trends", *Journal of Indian Museums* 57 (2004): 22–26.
8. Dan H. Nicolson, C. Suresh and Kattungal Manilal, *An Interpretation of Van Rheede's Hortus Malabaricus* (Königstein: Koeltz Scientific Books, 1988).
9. Ray Desmond, *The European Discovery of the Indian Flora* (Oxford: Oxford University Press, 1992). See also A. Galletti, A. J. van der Burg and P. Groot, *The Dutch in Malabar. Selections from the Records of the Madras Government, Dutch Records No. 13, Being a Translation of Selections No. 1 and 2 with Introduction and Notes* (Madras: Government Press, 1911).
10. Kattungal S. Manilal, "The Portuguese and Dutch plant explorations in 16th and 17th century Malabar and their socio-cultural and economic impact", *Journal of South Indian History* 2 (2005): 41–65. See also Kattungal S. Manilal, "The relationship of *Hortus Malabaricus* with Kerala and Malayalam", *Malayalam Literary Survey* 3 (1979): 54–67.
11. *Ibid.* See also Kattungal S. Manilal (ed.), *Botany and History of* Hortus Malabaricus (New Delhi: Oxford and IBH Publishing Company, 1980).
12. Kattungal S. Manilal, "*Hortus Malabaricus*, a book on the plants of Malabar and its impact on the religions of Christianity and Hinduism in the 17th century Kerala", *Indian Journal of Botanical Research* 1 (2005): 13–28.
13. *Ibid.*
14. Itty Achuden's caste was recorded in all four certificates printed in *Hortus Malabaricus.*
15. Kattungal S. Manilal, "The Portuguese and Dutch plant explorations in 16th and 17th century Malabar and their socio-cultural and economic impact".
16. Kattungal S. Manilal, C. Satish Kumar and M. Remesh, "Carl Linnaeus and *Hortus Malabaricus*: a 250th anniversary tribute to *Species Plantarum*", *Rheedea* 13, nos. 1 and 2 (2003): 3–18.
17. Richard H. Grove, "Indigenous Knowledge and the Significance of South-West India for Portuguese and Dutch Constructions of Tropical Nature",

in *Nature and the Orient: The Environmental History of South and Southeast Asia*, eds. Richard H. Grove, Vinita Damodaran and Saptal Sangwan (Oxford: Oxford University Press, 1998). Originally published in *Modern Asian Studies* 30 (1996): 121–143.

18. *Ibid.*
19. *Ibid.*
20. The other was *Hortus Elthamensis,* 1732 by Dillenius.
21. Kattungal S. Manilal, C. Satishkumar and M. Remesh, "Carl Linnaeus and *Hortus Malabaricus*: a 250th anniversary tribute to *Species Plantarum*".
22. Michel Adanson, *Familles des Plantes.* II. Partie (Chez Vincent, Comte de Provence 1763).
23. Kattungal S. Manilal, C. Satishkumar and M. Remesh, "Carl Linnaeus and *Hortus Malabaricus*: a 250th anniversary tribute to *Species Plantarum*".
24. Kattungal S. Manilal, *Van Rheede's Hortus Malabaricus. English Edition with Annotations and Modern Botanical Nomenclature* (Kerala: University of Kerala, 2003).
25. *Ibid.*
26. Annamma Spudich, *Such Treasure and Rich Merchandize: Indian Botanical Knowledge in 16th and 17th Century European Books* (Bangalore: The National Centre for Biological Sciences, 2008).

6 On Nature's Terms

Preserving the Practice of Traditional Backcountry Recreation in New Zealand's National Parks

Lee Davidson

Over time, New Zealanders have both shaped and been shaped by nature. Their evolving relationship with natural heritage has been responsible for important aspects of the national character and identity, forged through responses to a distinctive landscape. This chapter focuses on protected natural landscapes as natural heritage and the human engagement with these landscapes through the practice of mountaineering in New Zealand. The history of this engagement traces the development of a tradition of 'backcountry' recreation, which is both a physical and a cultural response to the natural landscape. This tradition encapsulates values such as self-reliance, a preference for encountering natural landscapes 'on nature's terms' and the experience of remote places as the reward for physical effort. These values and the experiences that engender and inspire them depend upon the protection of particular characteristics and attributes of natural landscapes, including 'natural quiet' and the absence of certain human technologies and modifications.

If, as has been argued, all heritage is socially constructed and involves an ongoing "negotiation of identity, place, and memory" through action and "cultural performance", then distinguishing between different types of heritage based on their materiality or immateriality is somewhat fraught.[1] Smith[2] makes the claim that all heritage is intangible so as to challenge the privileged status of the material in traditional Western heritage discourse. However far the argument is taken, it is clear that we cannot have one without the other, as heritage is not heritage if it is divorced from the values and meanings we attach to it. Calls to consider the "subtle interplay between tangible and intangible elements" of heritage and to understand it as a dialogical rather than an oppositional relationship seem well-founded.[3] In this respect the tradition of mountaineering in New Zealand is a good illustration of heritage as "a cultural process of engaging and experiencing" and a "multilayered performance" and the importance of heritage lying not just in a particular landscape but in what is done with it.[4]

Intangibility is associated with the practice and the creative acts of heritage, which constantly evolve through renewal and transmission, and are integral to a sense of identity, continuity and belonging.[5] However, in discussions of intangible cultural heritage, it has been stressed that the compartmentalization of the tangible and the intangible risks seeing one as dead, static and preserved, and the other as 'living'.[6] In the example discussed in this chapter, the *tangible* natural heritage is the geological forms, flora and fauna and ecological processes contained within the protected areas. The corresponding *intangible* natural heritage can be seen as the "interpenetration of human and natural spaces and processes".[7] Both are 'living,' and it is their dynamic interaction, which constitutes them as heritage.

This chapter looks in detail at one, quite specific form this interpenetration has taken in New Zealand—the development of a recreational tradition, which is underpinned by, and continues to depend upon, a particular state of natural heritage. The issues are further examined through a case study of one area within Fiordland National Park in New Zealand, which is being managed with the aim of preserving these attributes and allowing for continued opportunities to practice traditional backcountry alpinism. The case study illustrates the value of understanding recreation traditions as part of the intangible heritage of protected natural landscapes and demonstrates the importance of considering the interrelationships between tangible and intangible natural heritage for protected area management. Another consideration is how these traditions are communicated and interpreted for a wider public and whether or not there are also broader possibilities for promoting an 'environmentally benign culture.'[8] I will first discuss the relationship between culture and nature, as it relates to the historical development of protecting natural landscapes. I will then trace the history of mountaineering in New Zealand and its continued legacy as part of the tradition and values expressed in the contemporary practice of backcountry recreation.

TOWARDS A 'CULTURE OF NATURE'

Just as the intangible stands in relation to the tangible, we cannot consider nature without culture. Again the relationship is dialogical: "Culture, of course, is natural, as nature is cultural".[9] The ways in which people see the environment as natural or unnatural is "complex, diverse, overlapping and contradictory" and frequently "a site of struggle and contestation".[10]

Due to ecological processes and human impact, the 'content' of natural heritage is constantly in flux,[11] as is our perception of it, which is based on our understanding of the past and changing social and cultural contexts. In this way, the "reading" of nature is "a cultural process", whether we see it as "an expressive realm of purity and moral power"; the "object of spectacle, beauty or the sublime; [or] as a recreational space to be roamed across".[12] In Western culture, the tradition of the sublime transformed what had previously been frightening into 'a meaningful aesthetic experience.' By the nineteenth century, Romanticism had nurtured an attraction to hot springs, mountains and wilderness that encouraged us to view nature as a site of "leisure and pleasure".[13]

As our cultural perceptions of nature changed, the processes of industrialization and urbanization accelerated the division between society and nature. With the establishment of national parks, beginning in the U.S. in 1872 and soon following in Canada, Australia and New Zealand, this division increasingly took on a spatial form, with nature taking up its position on the margins.[14] Thermal areas such as Yellowstone were the first to attract the attention of governments, due to concerns that these areas would be "despoiled by rapacious private exploitation".[15] Instead, it was felt they should be developed as tourist resources under government control and 'in the national interest.' Another influential factor in the establishment of early national parks was the availability of large areas of publicly owned land considered 'worthless' for any other commercial activity.

By the 1880s and 1890s, there was a growing acceptance in New Zealand of aesthetic values, tourism and the protection of natural assets for future generations, as rationales for the establishment of protected areas. New Zealand's first national park, Tongariro, was the result of a gift by Maori chief Te Heuheu Tukino in 1887. Te Heuheu's motives, Shultis believes, were to deflect other tribes in the area from laying claim to the area and to protect the extremely sacred nature of the three peaks to the Ngati Tuwharetoa from commercial exploitation. In parliamentary debates on the matter, tourism potential was a particularly persuasive argument as it rendered otherwise 'unproductive scenery' into an additional resource to add to the wealth of the young colony, as was concern for protecting natural scenic areas from 'inappropriate' development.[16]

At the same time, requests were being made by scientific societies and concerned natural historians for the protection of New Zealand's indigenous flora and fauna.[17] Species preservation was given consideration in the 1890s with the establishment of native bird sanctuaries, beginning with Resolution Island in 1891. The Government's leading botanist of the time, Leonard Cockayne undertook a survey of the central North Island which had a major influence in deciding the scope of Tongariro National Park. Shultis suggests that the setting aside of national parks was also a reaction against the strong class system and the total private ownership of the national estate in England. In establishing a distinctiveness from England, national parks in New Zealand also appear to have been a product of a growing sense of identity with the indigenous landscape.

Today, roughly one-third of New Zealand's total surface area is conservation land; that is, more than eight million hectares, almost half of which is formally protected with the status of a reserve or national park. It has been estimated that less than half of one percent of this land has been protected in preference to a use forgone. Consequently alpine environments are well represented, in contrast to areas such as tussock grasslands, wetlands and lowland forests, which in the past had "ready development potential".[18] Since 1984 a Protected Natural Area Programme (PNAP) has sought to address this imbalance by identifying key areas of ecological importance for inclusion in a comprehensive and representative network of protected ecosystems. Other key challenges for the Department of Conservation (DOC), who manages these areas, include biodiversity loss and the eradication of invasive, introduced species of flora and fauna.

In addition to its conservation management responsibilities, DOC is charged, under the Conservation Act 1987:[19]

- To prepare, provide, disseminate, promote, and publicize educational and promotional material relating to conservation; and
- To the extent that the use of any natural or historic resource for recreation or tourism is not inconsistent with its conservation, to foster the use of natural and historical resources for recreation, and to allow for their use for tourism.

The present day conservation estate, then, bears the legacy of its historical evolution in its dual role of protecting natural areas and encouraging public recreation. Determining the conditions and extent to which to 'allow' for tourism in protected areas remains a constant dilemma.[20]

National parks have evolved out of the tradition of Western rationality and retain the heritage of anthropocentrism and a dualistic conception of the world.[21] Subsequent efforts have been made to recognize their 'intrinsic' value, but they are predominantly justified for their benefits to humankind, mostly in ecological and economic terms,[22] and defined in terms of compatible and incompatible human use. The cultural nature of natural heritage is sometimes used as a basis to critique the notion of protecting 'unmodified' natural landscapes in areas such as national parks. In spite of arguments that nature is an "arbitrary, cultural construct" and that there is no 'wild' nature—that is, unmodified by humans—left for us to protect, Heyd[23] argues that the value of natural heritage nonetheless exists in part in juxtaposition with landscapes more obviously transformed by human activity, and this makes "opportune a reassessment of the relation between nature and culture as they come to bear on the notion of natural heritage".

In this context, Heyd argues that the term "nature" can play an important role in designating areas where the "spontaneity of nature" takes precedence over human spontaneity. That nature may sometimes prove impervious to human domination provides a moral lesson that challenges our assumed superiority and points towards the possibility of a more balanced relationship between nature and culture.[24] Heyd sees a "culture of nature" as a possible basis on which to build an "environmentally benign culture".

Despite the irretrievably cultural dimension of protecting natural areas, Heyd argues that in designating reserves or national parks we may act on nature without increasing its artificiality. We may, instead, "allow for a *freer* development of the spontaneity of nature within those boundaries than would otherwise take place".[25] What must be considered is how this can be done so as to respect the spontaneity of non-human nature and the expression of its specific qualities. What makes a particular site more natural than cultural "consists in the value that one attempts to protect and enhance: the spontaneity of nature *vs.* the spontaneity of human beings". Furthermore, a "nature-affirming culture of nature" requires that appropriate attitudes towards nature are fostered in those who live, work or play in and near protected areas, so that this culture may permeate society more widely.[26]

Heyd's proposition that protected natural areas be considered part of a "culture of nature", underpinned by a moral dimension, which gives preference to the "spontaneity of nature", circumvents the stalemated culture versus nature debate. It is, therefore, a useful context within which to consider the evolution of a recreation tradition based on interactions between nature and culture in remote alpine areas.

NEW ZEALAND'S MOUNTAINEERING TRADITION: HISTORY AND CURRENT PRACTICE

European settlers arriving in New Zealand during the 19th century were primarily concerned with 'taming' the land, and wilderness was viewed by many as an "idle waste".[27] The dream of creating a 'Pastoral Paradise' in the South Pacific led to the destruction of much of the natural character of the land and, according to Jones,[28] effectively alienated the settlers from it. A number of naturalists and explorers, however, recognized the qualities of wild landscapes, and flora and fauna, undisturbed by human impact, and by the late 19th century a few New Zealanders had become interested in mountains for 'sporting' reasons. Although they were initially influenced by the Victorian mountaineering tradition, it was soon felt that the practice in New Zealand would be distinctive from 'Home' and that its pursuit could be instrumental in forming and strengthening the character of the young colonials. Its subsequent development was both a response to the natural landscape and in dynamic interaction with other evolving cultural responses to nature as New Zealand moved from being a colonial to a post-colonial society.

Europeans had become interested in climbing mountains against the backdrop of the Scientific Revolution and the rise of the Romantic tradition, which allowed the mountains to be 'imagined' differently from before: as geological formations and places in which to encounter the sublime, rather than as frightening places to be avoided.[29] During the 18th century, a small number of individuals were climbing in the European Alps, and, by 1850, a few major ascents had been made by continental mountaineers. In the latter half of the 19th century, the English came to the forefront of mountaineering in Europe. The new 'sport' had a particular appeal for urban middle class culture in mid-Victorian Britain, due to its combination of effort and beauty. To the Victorian mind, patriotism, athleticism and romanticism were served by climbing mountains, and character was forged through struggle, challenge, hardship, cooperative effort and the conquering of adversity.[30] British mountaineering literature of the 19th century centered on a discourse of reward for hard and resolute effort, manliness, physical and moral fitness, competition and mastery over nature.[31]

By 1880, having climbed all the major peaks in the Alps, the English and their continental guides took their techniques and experience and launched expeditions to mountain ranges in other parts of the world.[32] In 1882, the Reverend William Spotswood Green, a member of the English Alpine Club, came to New Zealand to make the first attempt to climb Aoraki/Mount Cook,[33] the country's highest mountain. Aspiring New Zealand mountaineers read Green's 1883 account of his adventures in 'the Antipodes',[34] which was packed with the curiosity of scientific discovery, the imperialism of exploration and the experience of the sublime. One of these readers was Malcolm Ross of Dunedin, who was inspired to organize his own expedition to climb a local peak. Typifying the 'pluck and daring', colonial ingenuity and self-reliance of early mountaineering exploits in New Zealand, ice axes were improvised from manuka saplings and the blades of sheep shears, while horseshoe nails provided extra friction for their boots.[35]

The formation of the New Zealand Alpine Club (NZAC) in 1891 formalized the association of a dispersed and loosely connected group of enthusiasts "who for sundry reasons love the mountains and love to climb them and explore their fastnesses".[36] Although in its early years the NZAC had a small membership and failed to excite wider public interest,[37] it did play a central role in establishing and molding the identity of the country's fledgling mountaineering culture. While the club was modeled on the London Alpine Club, it was felt that the unique New Zealand conditions "will always tend to develop a different type of climber". In particular it was believed that New Zealand climbers would be more hardy and self-sufficient due to the fact "that they, being without guides, are forced to learn and do many things for themselves which the English climber leaves to his guide and never learns".[38] Ross[39] predicted that climbing in turn would "not be without some influence in framing the character of Southern New Zealanders".

The development of mountaineering, however, was interrupted by World War I (WWI), and it was not until the inter-war period of the 20th century that climbing mountains for recreation was taken up by significant numbers of young people. This was in part due to increased mobility and leisure, which was prompting a growth of outdoor recreation culture throughout the Western World.[40] Although this social movement developed its own flavor in each particular country, it was often linked to the embracing of nature and the rejection of the 'ills' of modern life and associated with the emergence of social and political movements concerned with the protection of wilderness. Tramping, skiing and climbing clubs sprouted up throughout New Zealand at this time and began developing mountains as recreational areas by building huts, cutting tracks and encouraging participation on a mass scale. The government supported these measures, and introduced its own initiatives to promote outdoor activities as a wholesome and inexpensive way to keep the populace healthy.

In New Zealand, this outdoor culture can also be linked to the growing assertion of a distinctive national identity. In the cultural scene this was expressed in a desire to discard all that was British and Victorian and to find something that could take its place. Artists and writers explored new ways of seeing and relating to the landscape,[41] at the same time as New Zealanders were taking to the hills in a physical search for familiarity with and attachment to nature.

This historical period was also one of dislocation and disillusionment, as well as expanding bureaucracies and rationalization through which New Zealand was losing much of the personal and informal flavor that had characterized it as a pioneering, frontier society.[42] The growing enthusiasm for mountain recreation can therefore be seen as a desire to inhabit, however temporarily, landscapes in which the values nurtured in the young colony could be maintained. These values were encapsulated in a simple and informal environment where what mattered most was character, companionship, 'kiwi ingenuity' and physical endurance.[43]

Mountaineers of this era prided themselves on their easygoing outlooks and 'give it a go' attitudes, making up for their lack of financial resources by fashioning technical equipment from whatever came to hand: packs made from used sugar bags and 'ice-axes' made from sticks cut from the bush (Russell Edwards personal communication, August 18, 1988). Following World War II (WWII), New Zealand mountaineering stepped onto the world stage with success in the Himalayas, culminating in two New Zealanders—Ed Hillary and George Lowe—being pivotal in the success of the British expedition, the first to climb Mount Everest in 1953. This was an important endorsement for the homegrown style of mountaineering, and the tradition and values of those earlier generations are still a crucial touchstone for the ethos of contemporary New Zealand mountaineers. As one of the first two people to stand on the summit of Everest, and by his subsequent adventures and charitable work, Hillary has become a central figure in the cultural myth of the New Zealand character based on resourcefulness, perseverance and humility.

I now draw on my recent in-depth study of New Zealand mountaineers using biographical narrative interviews.[44] The material is used here to illustrate the ongoing importance of mountaineering heritage in New Zealand as the basis of a tradition and a set of values that facilitate the construction of identity and meaning through engagement with alpine landscapes. Contemporary mountaineers view good style as "efficient, fast climbing",[45] just an "ice axe and ice hammer and crampons and, go for it". Climbing in this style—"strong and fast and get things done"—is something that New Zealand climbers pride themselves on and something that is particular to their identity and the heritage of an apprenticeship in the New Zealand mountains. As one climber put it, it is not "just being on the mountain":

> *It's the getting in and getting out, that side of it, and New Zealand climbers generally, they grow up having to do that . . . you do these big long, sort of missions and that was really satisfying.*[46]

Even today, the mountains in New Zealand are not as accessible as they are in other parts of the world, such as the European Alps, so the equipment taken has to be minimized, and speed maximizes the opportunities presented by short periods of settled weather in the climate of temperate islands. This style of climbing is less reliant on modern forms of technology, for example, helicopters or *téléphériques*, or less modern labor, such as porters and Sherpas, in order to get climbers and their equipment to and from climbs. This makes it a more self-sufficient and economical exercise. Going 'on nature's terms' means that ingenuity, speed, efficiency and being able to read and adapt to the dynamics of the environment are valued over the use of technology to conquer and dominate nature.

Many climbers express regret that not as much of this style of climbing is being done anymore. There is seen to be a trend away from traditional New Zealand alpine climbing, with people not being as prepared to extend their boundaries by undertaking 'big long missions,' walking in and out, carrying a heavy pack, and exploring remote country. People who climb in the traditional way are contrasted with those who are viewed as not having the depth of commitment to be self-reliant; for example, guided climbers, or those who fly in and out of climbing areas rather than taking the time to walk in. Heli-climbers, as one climber sees it, "crave the instant mountain gratification experience".[47]

Similarly, those who consume the products of the 'adventure industry' are "people who want their creature comforts, and who aren't prepared to go on nature's terms", and this is seen by committed mountaineers as having consequences for the kinds of experiences available to them:

> *. . . a lot of people engaging in this adventure tourism will do these things once and that's it, and boast about that until it sort of gets boring. It probably won't have any . . . major meaningful change or effect on their, on their character, on their strength, on their self reliance. . . . I do see a lot of really beautifully, physically, fit young people doing this stuff and yet they haven't got a clue really about what it's all about.*[48]

Mountaineers see the motivations here as being ego or status, as opposed to a "love of the mountains", which underpins the traditional style of mountaineering and the lifelong commitment it requires. The strong emotional attachment to the mountains that climbers profess is often traced back to their childhoods and described as "a wave of good feeling", "a euphoric sort of thing", feeling "at home" in the mountains and having "a little piece of joy that you carry around with you". These feelings give mountaineers a strong sense of coherence and continuity in their lives, which is supported by belonging to a community and a tradition of "like-minded" people.

Self-reliance is also central to this identity. Traveling through remote and 'wild' alpine landscapes means they have no 'crutches' or outside assistance; they must rely exclusively on their own skills and resources, and those of their companions. As one climber argued, the mountains should be approached "purely on nature's terms" so that "through the discovery of land . . . one can discover oneself . . . on the real terms that are not contrived":

> *That whole business of being able to equip yourself and go and live out there. And experience what that means. And that's the guts of it. That's the raw guts of what it's about really. We live such cloistered lives.*[49]

The committing situations mountaineers encounter in remote locations also provide the opportunity to grow, as they sometimes have to "push" their boundaries and thereby discover abilities they did not previously know they possessed. The challenges this environment offers are seen as an "honest test", because there is immediate and unambiguous feedback about their abilities: "if you make a mistake you pay for it". In this context there is no room for ego—you have to have a realistic sense of your abilities, as to overestimate them could have serious consequences.

The values that are seen as essential to being a "true mountaineer"—commitment, love of the mountains, humility, simplicity and responsibility (for one's companions and the environment)—provide a sense of 'solidity' in a world, which, as Bauman[50] puts it, "seems less solid than it used to be". They form a moral framework, which is seen as arising from intrinsic aspects of the environment itself and the experience of direct engagement with these. Contributing to this sense of solidity is, first, a *clarity* and *sense of control* that differentiates mountaineering from everyday life and a sense of *perspective* that can be a guide for life beyond the mountains.

On a climbing trip, mountaineers become "completely involved" in what they are doing and feel they are in their "own little world" that is "totally removed from day to day life". Having to focus on the immediate present and block out distracting thoughts and emotions, simplifies the world of mountaineering and differentiates it from the world of the everyday. It is an absorbing, uncomplicated way of being that gives them a sense of being able to think more clearly. They feel more aware of who they are with and what is going on around them. The decisions they face are "black and white". Even when things go wrong, it is generally in a clear and unequivocal way. That climbing requires "self control", sometimes "for days on end", can be a source of strong feelings of competence that arise from:

> *that feeling of total aloneness, just us, and that complete, emptiness all around you. But I loved the storms and things as well because we*

> *knew how to cope with bad weather . . . and it never felt, endangered. . . . we had some amazing experiences that didn't come off the way they were meant to come off, but that didn't actually matter, as long as we got ourselves out of our predicament [laughs].*[51]

Experiences in the mountains are seen to represent and lead one towards what is "real" and "important"; that is, this environment provides a clearer, less ambiguous moral source. Familiarity with the "solid" mountains and the values of mountaineering made people "happier" and "nicer" and their lives "easier", "simpler and "more sane", by changing their focus and putting things in "perspective". While the world of the everyday is "frivol[ous]", "silly", "trivial", "the trappings", "contrived", the world of mountaineering is "real", "down to earth", "basic" and, again, "simpler". It helps them distinguish what is trivial from what is important in life:

> *Nothing compared to putting your life out there on a bit of a line at times climbing so . . . you'd go into the mountains and that was reality, back out, back at home, back at work or varsity, that was just living you know and that put everything in perspective in a lot of ways, it made life quite easy I reckon. You didn't seem to get caught up in, it just seemed to focus you on, on what was real in life, what was important in life and what wasn't. . . . you can look back through your life and there's key decisions you make that take you one way or another way. I think they're relatively easy to make. . . . the living as opposed to the trappings . . . you look at some people just totally focused on, on possessions. . . . I think [mountains] make you quite solid.*[52]

In summary, the traditional New Zealand mountaineering experience and identity has evolved out of the realities of a rugged and remote landscape and a heritage based on self-reliance and 'colonial ingenuity,' which, it was believed, would distinguish local climbers from the Victorian gentleman mountaineer and help shape the New Zealand character. These ideals were strengthened throughout the dislocation and hardship of the interwar period and were, in a sense, vindicated in the 1950s with Hillary and Lowe's success on Everest. These values remain central to the contemporary New Zealand backcountry experience, although there are concerns that they are threatened by societal change and the impact of commercialism and new technologies. The enduring benefits of this tradition derive from the implications that self-reliance have for gains in self-knowledge, humility and respect for the environment, as well as a sense of clarity, competence and perspective. To go 'on nature's terms' therefore implies a particular kind of experience, related to New Zealand's alpine heritage and a moral relationship with nature embodied within it.

CASE STUDY: MANAGING 'REMOTE EXPERIENCES' IN NEW ZEALAND'S NATIONAL PARKS

The tradition of mountaineering in New Zealand and its associated values and practices are part of the intangible natural heritage of alpine national parks in New Zealand. The need to manage national parks in a way which allows for the preservation of this intangible heritage—through its continued practice—can be highlighted by a case study of Fiordland National Park, and the conflicts surrounding a recent review of its management plan. It also demonstrates the ways in which perceptions of natural heritage are contested, and the need to understand the basis and implications of different perceptions in order to make sound management decisions.

Fiordland National Park

Fiordland National Park (FNP), covering an area of 1,260,200 hectares in the southwest corner of the South Island, is the largest of the 14 national parks in New Zealand and one of the largest in the world. It is part of Te Wahipounamu—*South West New Zealand* World Heritage Area and receives around 500,000 visitors annually. The FNP was given national park status in 1952 on account of its "impressive natural attributes" and appeal for visitors,[53] which are described in its Management Plan:

> *Fiordland is renowned for its combination of glaciated mountain landforms, remote coastline indented by fiords, vast extent of natural flora, and diverse and abundant wildlife.... Glaciers of past ages have carved out the sheer and rugged landform. Water now permeates the landscape in the form of lakes, rivers and ocean; also snowfields and rain.... [Fiordland] is the largest area of very strong crystalline rocks in New Zealand . . . The Fiordland coast, with mountains rising up to 2000m directly from the sea along the Alpine Fault, forms a barrier to strong westerly winds, which travel across the Southern Ocean and Tasman Sea. The abrupt obstruction causes a very wet, vigorous climate. . . . Irregular ridges and valleys, waterfalls that plunge down the steep mountain walls, countless hidden lakes and tarns, successive marine benches (or terraces) on the south coast, and the mixture of U- and V-shaped valleys resulting from glacial and river erosion, all contribute to the distinctive Fiordland landscapes. The valleys and lower mountain slopes are clothed in temperate rainforest while the upper slopes rise through alpine grasslands and herbfields to high peaks, some retaining small glaciers. . . . The lakes and rivers are scenic gems enhanced by the grandeur of indigenous forests and towering snow-capped mountains. Visitors' impressions of Fiordland are the almost overpowering steepness of the mountains, the comparative narrowness of the valleys and inundation by water. The presence of water and*

> *its influence predominate throughout Fiordland National Park, providing a wide range of ecological, scenic and recreational attributes.*[54]

The first humans to traverse this landscape were the Maori, who came in search of pounamu (greenstone/nephrite), birds, fish and shellfish.[55] The FNP is also important in relation to the history of European exploration of New Zealand and early industry including sealing and whaling, gold mining and sawmilling. However, the rugged nature of the terrain made attempts to establish a port and pastoral settlement a failure.[56] While the potential for resource exploitation was being explored in the late 19th century, many of the islands along the Fiordland coast were designated as reserves and pioneering wildlife conservation efforts began. The area of Milford Sound was settled by Europeans in the 1870s and soon became a tourist attraction, and it remains one of New Zealand's premier tourist sites. The establishment of the Milford Track from 1888 provided another means by which tourists could experience the natural beauty of the area, and this multi-day trek continues to attract walkers from around the world.[57] Other popular tourist and recreational activities in the park include boat trips, fishing and hunting, sightseeing and scenic flights, while the granite peaks of the Darran Mountains provide some of the best alpine rock climbing in New Zealand. Road access into Milford Sound was finally established with the completion of the Homer Tunnel in 1953, but a large number of tourists take the opportunity of a shorter, scenic flight from the tourist hub of Queenstown.

Management of FNP

The DOC manages Fiordland National Park in accordance with the National Parks Act 1980. Under the Act, national parks are to be preserved "for their intrinsic worth and for the benefit, use, and enjoyment of the public".[58] The 2005 General Policy for National Parks[59] provides a "consistent national direction" for the implementation of the Act and individual park Management Plans must be consistent with the General Policy. Throughout this policy there is frequent mention that the enjoyment of national parks "should be on nature's terms", and it recognizes them as "places of refuge from commercialism and urbanization" where people can find "natural quiet" and "release from the stresses of everyday life".[60] There is also recognition of the role of national parks in reinforcing a national identity through 'the opportunity to see, hear or feel natural New Zealand' and that the "traditional New Zealand back country experience has a distinctive character . . . [which] requires self-reliance and its focus is on nature being met on nature's terms".[61]

A major challenge for the DOC in recent years has been significant growth in the number of people visiting national parks, particularly international visitors, and changing recreation patterns.[62] Marketed for its qualities of outstanding natural beauty and relatively 'unpeopled' landscapes, international tourism to New Zealand between 1985 and 1995 rose dramatically from 670,000 to 1.4 million. Tourism arrivals in New Zealand continue to grow, reaching 2.5 million in the year ending August 2010.[63] Research conducted by the Ministry of Tourism suggests that over half of these tourists visit some part of the public conservation estate, either for picnicking, short walks, scenic drives (to Milford Sound, for example) or on longer tramping or climbing trips.[64] This has led to concern among outdoor recreation groups that traditional backcountry recreation is being compromised by rapidly growing tourist demand for nature-based experiences resulting in overcrowding and environmental degradation of popular areas. There is also concern that DOC will cater for this tourist demand by providing higher-grade facilities in the 'frontcountry,' and pricing these facilities out of the range of many New Zealanders, while neglecting the 'backcountry' areas used by traditional recreationists.[65]

Another contentious issue has been overflights by aircraft of national parks and the use of aircraft to access remote locations. The estimated number of tourists taking scenic flights is 50,000 to 100,000 annually nationwide.[66] The noise of aircraft in national parks has been found to detract from the enjoyment of the backcountry recreation experience.[67] This issue arose in a recent review of the Fiordland National Park Management Plan, when the DOC sought to restrict the number of scenic flights in and out of Milford Sound[68] and revise the regulations for landing elsewhere in FNP.

Various parts of FNP are zoned for specific conservation purposes and a range of recreational experiences (Figure 6.1). The five visitor management settings are wilderness areas; remote experience; backcountry; high use track corridors and frontcountry. These are not designated formally in legislation, but are managed by the facilities that are provided (or not) and the restriction of aircraft access.

Within the park, lying between the Hollyford Track and the Milford Road, is the Darrans Remote Area (DRA), recognized as an area valued for 'premium alpine and hard rock climbing.' Remoteness seekers as a visitor group are regarded as being "self-reliant trampers, hunters and mountaineers who want a true wilderness experience".[69] The plan review sought to manage opportunities for remote experiences in the DRA "with minimum disturbance from aircraft", and proposed that only limited aircraft access be permitted at designated landing sites. However, the FNP Draft Management Plan, first released for public comment in 2002, proposed doubling the number of landing sites from two (Turner's Bivvy and Ngapunatoru Plateau) to four (with the addition of Te Puoho Glacier and Lake Turner; Figure 6.1), thereby extending access into the Central Darrans. It was also proposed to make access seasonal (May 1 to November 30) and for climbing and ski touring only, that is, no scenic landings except on Ngapunatoru Plateau where there was to be a maximum of six landings per day, year-round, including glacier visits for tourists.

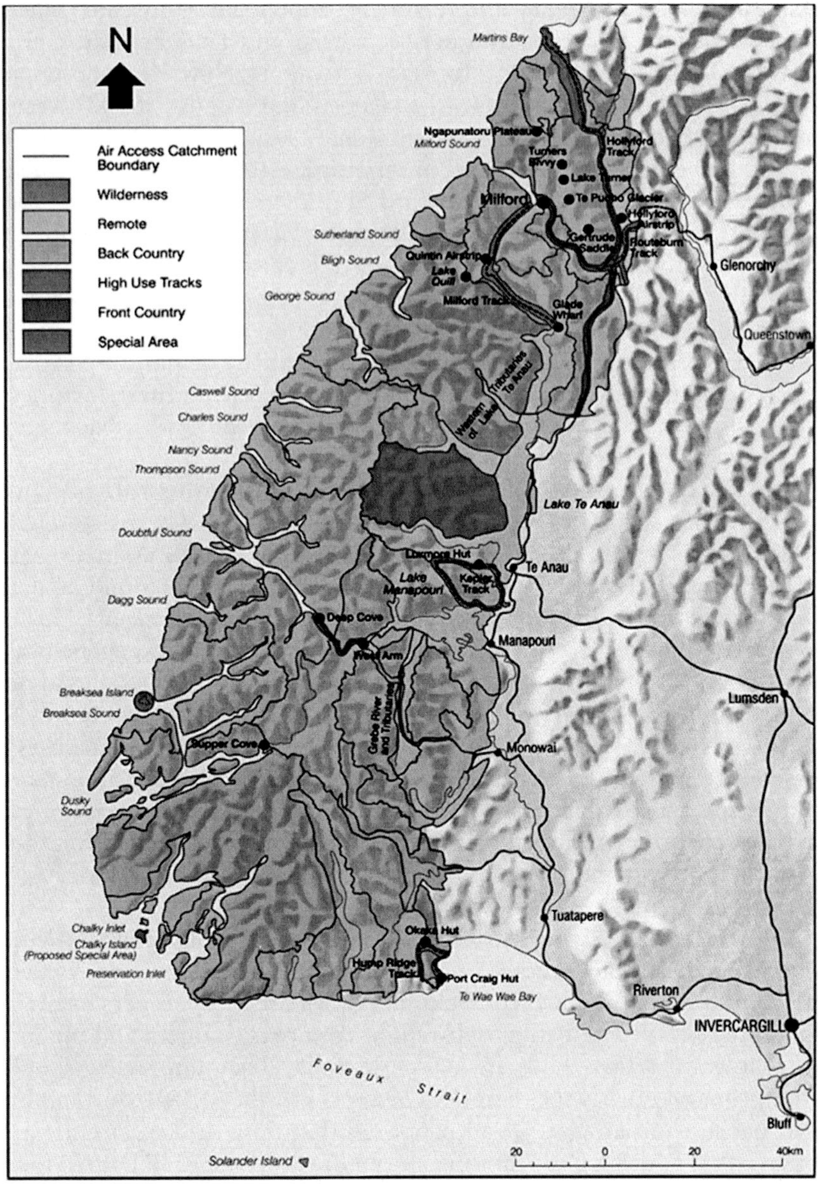

Figure 6.1 Proposed air access sites, Draft Fiordland National Park Management Plan, 2002, page 110. Image reproduced courtesy of the DOC, New Zealand.

The response to these proposals from the principal active climbers in the area was a "near unanimous and strong objection".[70] The arguments expressed in public submissions against extending air access into this remote area included the concern that it would threaten the special qualities of the area, in particular its difficult access, feelings of isolation, its status as an 'aspirational area' for climbers, its relatively untouched and unexplored nature and the current focus on 'traditional' climbing styles in the Darrans (Figure 6.2). Practical considerations were also mentioned, such as an increase in rubbish and human waste accumulating in fragile settings, and fly-in parties being not as self-sufficient and therefore potentially unable to walk out on their own should inclement weather prevent them flying out. It was also felt that increasing air access would set a precedent, becoming 'the thin edge of the wedge' for a future loosening of restrictions. Other arguments encapsulated moral considerations drawn from the mountaineering tradition: that to walk into this area rather than fly was to go 'on the mountains' terms' and thereby with the requisite respect and that the opportunity to enjoy it should be the reward for the effort and commitment required to get there. Some submitters argued that by maintaining these conditions –that is, by disallowing recent technologies of access –the heritage and traditions of climbing in New Zealand would be preserved. The NZAC submission opposed additional landing sites but supported making air access seasonal and extending this from March 1 to November 30. This would prohibit landings during the best part of the climbing season—the southern hemisphere summer—while accommodating those who wanted to fly in at the end of the season (March and April).

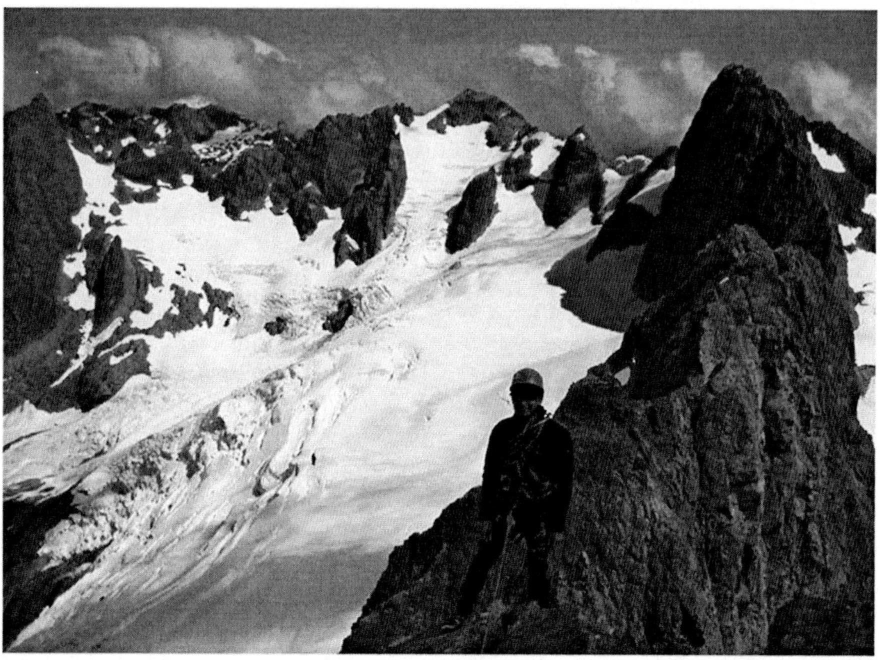

Figure 6.2 Climber on Te Wera, Central Darrans, with Te Puoho Glacier behind. Photo courtesy of John McCallum.

Those who supported increased air access into the DRA argued that it catered for the less fit and able and the time-pressured, who would otherwise be unable to visit. In response to concerns about disturbance from aircraft noise, it was argued that scenic overflights of the area caused the most impact, and additional landings would therefore be insignificant in terms of disrupting 'natural quiet'. Air operators opposed any restrictions on aircraft landings. The final wording of the FNP adopted the recommendations of the NZAC, thereby accepting the preferences of 'traditional' climbers in relation to access conditions in this one area of FNP. However, debates continue in other alpine national parks in New Zealand regarding issues of motorized access and the control of commercial activities in the interests of recreational backcountry experiences.

CONCLUSIONS

The preservation of intangible heritage depends upon its ongoing practice by which the meanings and values that it represents are kept alive. This ensures the continuity from the past to the present, maintaining a sense of belonging and of place upon which identity and moral relationships are built. I have argued that a traditional recreation practice—mountaineering—is an integral part of protected natural heritage in New Zealand. Its history is intertwined with the history of New Zealand's national parks and it is an embodiment of an important aspect of the evolving 'interpenetration' of people and nature in this country. In particular, through traditional climbing styles, New Zealand mountaineers 'perform' moral values which are part of their heritage and identity. These moral values imply limits to the human technologies, which should be used in remote landscapes, so as to allow for an experience in which natural processes and conditions are privileged in a way that differs significantly from 'everyday' life.

These experiences have important personal benefits for the individuals involved, but they also have wider cultural importance and other 'intangible' benefits which could help to inform the debate around heritage conflicts of this nature. The sense of loss on the death of Hillary in 2008, and the ongoing need in New Zealand to celebrate and cement his place in the cultural imagination, signifies the importance of keeping alive the climbing tradition of which he was a product, especially in times of rapid social change. This requires the management of some alpine areas in New Zealand in a way that protects them from the intrusion of certain technological and economic processes. In the public debate that surrounds the use of protected natural areas, recognition of the cultural importance of this tradition would serve as a valuable counter argument to the view that such restrictions 'lock up' natural resources for the benefit of an elite group of individuals at the expense of commercial interests.

In this case study, through 'artificial' management decisions, the intangible natural heritage specific to New Zealand mountaineers has been preserved, in a place which holds significant meaning for them. But it also demonstrates the way in which intangible heritage values may not be well understood by the wider public, and can come under pressure—in this case from the tourism industry and aircraft operators in particular. In light of this, an argument could be made for more interpretation of natural heritage such as alpine landscapes in national parks, which includes these types of intangible heritage, and explains their historical basis and contemporary relevance. If the cultural legacies of intangible natural heritage are widely communicated to the public, they are less likely to be lost amidst the power of arguments given weight by tangible economic benefits.

Finally, the existence and continued ecological integrity of alpine national parks is the first condition for the ongoing practice of traditional mountaineering in New Zealand. In this sense materiality is essential to the 'performance' of heritage by mountaineers engaged in traditional climbing styles. But the interplay between the tangible and intangible dimensions of this natural heritage could also have an element of reciprocity. The principle of 'going on nature's terms' embodied in this tradition could act as a vital moral lesson about the intangible benefits of respecting nature and its spontaneity and accepting limits to human behavior that preserve this. If this heritage is kept alive and its meanings shared with a wider public, it could be part of current social and cultural change, as we seek to build a more environmentally benign culture for the future. The argument that we all need to proceed 'on nature's terms' may be advanced by preserving both the places and practices in which the 'spontaneity of nature' takes precedence.

NOTES

1. Charlotte Andrews, Dacia Viejo-Rose, Britt Baillie and Benjamin Morris, "Conference report: tangible-intangible cultural heritage: a sustainable dichotomy? The 7th Annual Cambridge Heritage Seminar", *International Journal of Intangible Heritage* 2 (2007): 124–129.
2. Laurajane Smith, *Uses of Heritage* (New York: Routledge, 2006).
3. Charlotte Andrews, Dacia Viejo-Rose, Britt Baillie and Benjamin Morris, "Conference report: tangible-intangible cultural heritage: a sustainable dichotomy? The 7th Annual Cambridge Heritage Seminar".
4. Hugh Cheape, Mary-Cate Garden and Fiona McLean, "Editorial: heritage and the environment", *International Journal of Heritage Studies* 15 (2009): 104–107.
5. Richard Kurin, "Safeguarding intangible cultural heritage: key factors in implementing the 2003 Convention", *International Journal of Intangible Heritage* 2 (2007): 10–20. See also Laurajane Smith, *Uses of Heritage*.
6. Marilena Alivizatou, "Contextualising intangible cultural heritage in heritage studies and museology", *International Journal of Intangible Heritage* 3 (2008): 44–54.
7. Thomas Heyd, *Encountering Nature: Toward an Environmental Ethic of Care* (Burlington: Aldershot, 2007).

8. Thomas Heyd, *Encountering Nature: Toward an Environmental Ethic of Care.*
9. Peter Howard and Thymio Papayannis (eds.), *Natural Heritage: At the Interface of Nature and Culture* (New York: Routledge, 2007).
10. Phil Macnaghten and John Urry, *Contested Natures* (London: Sage, 1998).
11. Giorgos Catsadorakis, "The Conservation of Natural and Cultural Heritage in Europe and the Mediterranean: A Gordian Knot?" in *Natural heritage: At the Interface of Nature and Culture*, eds. Peter Howard and Thymio Papayannis (New York: Routledge, 2007).
12. Phil Macnaghten and John Urry, *Contested Natures.*
13. *Ibid.*
14. *Ibid.*
15. John Shultis, "Natural Environments, Wilderness and Protected Areas: An Analysis of Historical Western Attitudes and Utilisation, and Their Expression in Contemporary New Zealand" (PhD diss., University of Otago, 1991).
16. *Ibid.*
17. *Ibid.*
18. Jonathan G. Clease, *Towards Sustainable Ecotourism: Management and Recommendations* (Master's diss., University of Otago, 1994).
19. Conservation Act 1987, No. 65, accessed 26 May, 2011, http://www.legislation.govt.nz/act/public/1987/0065/latest/whole.html#dlm103610
20. See, for instance, Jenny Chamberlain, "Wild space race: loving the wilderness to death", *North and South* 203 February (2003): 48–61. See also Tara Ross, "Great Outdoors Become Great Rat Race", *Sunday Star-Times*, November 28, 2004.
21. Chris Rojek, *Ways of Escape: Modern Transformations in Leisure and Travel* (Basingstoke: Macmillan, 1993).
22. David Lowenthal, "Natural and Cultural Heritage", in *The Nature of Cultural Heritage and the Culture of Natural Heritage: Northern Perspectives on a Contested Patrimony*, eds. David Lowenthal and Kenneth Olwig (New York: Routledge, 2006).
23. Thomas Heyd, *Encountering Nature: Toward an Environmental Ethic of Care.*
24. *Ibid.*
25. *Ibid.*
26. *Ibid.*
27. Lawrence Jones, "Versions of the Dream: Literature and the Search for Identity", in *Culture and Identity in New Zealand*, eds. David Novitz and Bill Willmott (Wellington: GP Books, 1989) 187–211. See also Trudie McNaughton (ed.), *Countless Signs: The New Zealand Landscape in Literature* (Auckland: Reed Methuen, 1986).
28. Lawrence Jones, "Versions of the Dream: Literature and the Search for Identity".
29. Robert MacFarlane, *Mountains of the Mind: Adventures in Reaching the Summit* (New York: Vintage Books, 2004).
30. Peter Donnelly, "Climbing is character building: the second of four fallacies", *Mountain* 82 (1981): 20–23.
31. David Robbins, "Sport, hegemony and the middle class: the Victorian mountaineers", *Theory, Culture and Society* 4 (1987): 579–601.
32. Philip Temple, *The World at Their Feet* (Christchurch: Whitcombe and Tombs, 1969).
33. New Zealand's highest peak has two official names. 'Aoraki,' or the regional variation, 'Aorangi,' is the Māori name. Aoraki was given the name Mount

Cook by Europeans to commemorate the achievements of the English explorer James Cook.

34. William Spotswood Green, *The High Alps of New Zealand* (London: Macmillan, 1883).
35. Walter Scott Gilkison, *Earnslaw, Monarch of Wakatipu* (Christchurch: Whitcombe and Tombs, 1957). See also Malcolm Ross, "Mountaineering in Otago", *New Zealand Alpine Journal* 1 (1892): 30–37 and Malcolm Ross, *A Climber in New Zealand* (London: E. Arnold, 1914).
36. George E. Mannering, "Editorial" *New Zealand Alpine Journal* 1 (1892): 8–14.
37. George E. Mannering, "The inauguration of the New Zealand Alpine Club", *New Zealand Alpine Journal* 1 (1892): 3–5.
38. George E. Mannering, "Editorial".
39. Malcolm Ross, "Mountaineering in Otago".
40. Melissa Harper, *The Ways of the Bushwalker: On Foot in Australia* (Sydney: University of New South Wales Press, 2007). See also Kirstie Ross, *Going Bush: New Zealanders and Nature in the Twentieth Century* (Auckland: Auckland University Press, 2008).
41. Gordon H. Brown, *New Zealand Painting 1920–1940: Adaptation and Nationalism* (Wellington: Queen Elizabeth II Arts Council of New Zealand, 1975). See also Trudie McNaughton (ed.), *Countless Signs: The New Zealand Landscape in Literature.*
42. Erik Olssen, "Towards a New Society", in *The Oxford History of New Zealand*, ed. William H. Oliver (New York: Oxford University Press, 1981), pp. 250–278.
43. Lee Davidson, "The 'spirit of the hills': mountaineering in Northwest Otago, New Zealand 1882–1940", *Tourism Geographies* 4, no. 1 (2002): 44–61.
44. Lee Davidson, "A Mountain Feeling: The Narrative Construction of Meaning and Self through a Commitment to Mountaineering in Aotearoa/New Zealand" (PhD diss., Monash University, 2006).
45. Quotes are taken from the interview material in Davidson, "A Mountain Feeling: The Narrative Construction of Meaning and Self through a Commitment to Mountaineering in Aotearoa/New Zealand".
46. *Ibid.*
47. Brian Wilkins, "West Matukituki Valley and Bevan Col helicopter access", *The Climber* 1 (2004): 13–14.
48. Taken from Davidson, "A Mountain Feeling: The Narrative Construction of Meaning and Self through a Commitment to Mountaineering in Aotearoa/New Zealand".
49. *Ibid.*
50. Zygmunt Bauman, *Life in Fragments: Essays in Postmodern Morality* (Oxford: Blackwell, 1995).
51. Taken from Davidson, "A Mountain Feeling: The Narrative Construction of Meaning and Self through a Commitment to Mountaineering in Aotearoa/New Zealand".
52. *Ibid.*
53. DOC, *Fiordland National Park Management Plan* (Wellington, New Zealand: DOC, 2007).
54. *Ibid.*
55. *Ibid.*
56. *Ibid.*
57. In the period from October 2008 to April 2009, an estimated 14,400 people walked the Milford Track, accessed August 12, 2010, http://www.doc.govt.

nz/about-doc/role/maps-and-statistics/visitor-statistics-and-research/great-walks-visitor-statistics/.

58. Section 4(1), National Parks Act 1980, No. 66, accessed 26 May, 2011, http://www.legislation.govt.nz/act/public/1980/0066/latest/whole.html?search=ts_act_national+parks+act_resel&p=1#dlm36963.
59. DOC, *General Policy on National Parks* (Wellington, New Zealand: DOC, 2005).
60. *Ibid.*
61. *Ibid.*
62. Gordon Cessford and Paul Dingwall, "Wilderness and Recreation in New Zealand", in *The State of Wilderness in New Zealand*, ed. Gordon Cessford (Wellington, New Zealand: DOC, 2001), pp. 35–42.
63. Ministry of Economic Development, *Key Tourism Statistics* (Government of Wellington, New Zealand, October 1, 2010).
64. "Conservation" page of the Ministry of Economic Development, accessed October 10, 2010, http://www.tourism.govt.nz/Our-Work/Our-Work-Summary-page/Conservation/.
65. Lee Davidson "What Does It Mean to Manage Tourism Sustainably on New Zealand's Conservation Estate?" (Master's diss. (Applied), Victoria University of Wellington, 1997).
66. Alon Tal, "'Naturally quiet': towards a new legislative strategy for regulating air space above national parks in New Zealand", *Otago Law Review* 10 (2004): 537–574.
67. Gordon Cessford and Murray C. Reedy, "Wilderness Status and Associated Management Issues in New Zealand", in *The State of Wilderness in New Zealand*, ed. Gordon Cessford (Wellington, New Zealand: DOC, 2001), pp. 43–56. See also Michael Harbrow, *Effect of Air Traffic Associated with Milford Aerodrome on Visitors to Fiordland National Park: Summer 2007/08* (Invercargill, New Zealand: DOC, 2008). For further background, see Hollyford Track Survey, accessed February 23, 2011, http://www.doc.govt.nz/upload/documents/about-doc/role/hollyford-track-survey-report-2006–07.pdf and Key Summit Survey, accessed February 23, 2011, http://www.doc.govt.nz/upload/documents/about-doc/role/key-summit-day-visitor-survey.pdf.
68. Air traffic in Milford Sound is particularly heavy: "The annual number of aircraft movements (landings and take-offs) between 1996–2005 averaged about 17,000 (rounded to nearest 500). However due to unsuitable weather conditions these movements are generally limited to flying on approximately 200–250 days a year. Peaks of more than 200 movements per day have been recorded". DOC, *Fiordland National Park Management Plan* (Wellington, New Zealand: DOC, 2007).
69. For more on DOC's visitor settings, see the Visitor Strategy (1996), accessed February 24, 2011, http://www.doc.govt.nz/upload/documents/about-doc/role/policies-and-plans/visitor-strategy.pdf.
70. The author obtained summaries of submissions on the aircraft access section of the Draft Management Plan from the DOC under an Official Information Act Request. A copy of the full New Zealand Alpine Club submission was supplied on request by their Recreation Advocacy Convener.

7 Poetically, Man Dwells With Crickets

Nature and Culture of Chinese Singing Insects

Xingbao Jin and Alan L. Yen

INTRODUCTION

The highly successful 2010 Shanghai World Expo is now over. Its theme, "Better City, Better Life", has catalyzed the concept of 'eco-city', a human settlement that enables its residents to attain a high standard of living, while minimizing expenditure of natural resources.[1] For many rapidly developing cities, it is important for governors, planners, builders and inhabitants to understand the core philosophy of the eco-city, that is, to rebuild or improve existing cities to attain a balance with nature based on principles for long-term ecological sustainability, cultural vitality and the health of the Earth's biosphere. This concept has to be embodied in many aspects of the city management: the buildings, the transport system, industry, the economy, cultural vitality, the state of the urban natural environment and biodiversity.

The natural environment is, to a great degree, affected and shaped by human culture, in turn the result of social activities.[2] Understanding cultural values that have a positive influence on human behavior is essential to contributing to public awareness of, and sensitivity to, nature within the urban environment, as well as to create an understanding of the need for biodiversity conservation. Ultimately, the scientific research and science communication produced by natural history museums contributes to enabling people to care for their local environment and to become sensitive to its aesthetic and evocative qualities.

In this chapter, we explore the natural history of singing insects in the context of Chinese culture and as a case study for raising awareness of urban biodiversity. How does this ancient tradition influence popular pastimes today? What is the importance of these common orthopteran insects to the urban environment? Can the "Cultural/informational web"[3] be used for nature conservation and science communication for a better understanding of the criteria required for a better city or an eco-city?

THE ANCIENT CULTURE ALIVE

There is a typical Chinese ink painting from the South Song Dynasty (AD 1127–1279) deposited in the Metropolitan Museum of Art, New York, entitled "*Bin Feng July*" (《豳风 七月》), meaning "the melody of July", which is a painting based on the content of the first Chinese poem collection, the Book of Odes, "*Shi Jin*" (《诗经》), 11th BC–AD 6th).[4] The whole painting reel is over 16 m long, a width of about 30 cm high, with 34 pictures, of which three are about the singing insects described in the poem:

> [They are] *In the wild in July,* (七月在野，)
> *Below the eaves in August,* （八月入宇,）
> *Inside the house in September,* （九月在户,）
> *And under my bed in October.* （十月蟋蟀入我床下.）

The painting depicts 10 katydids with broad wings, long antenna and strong hind legs, jumping into the yard from the field, into the house, finally into the dream of the poet, who is lying on his bed, listening to the melody sung by these little creatures and apparently sighing with emotion. It is the same emotion evoked in "Cricket" by the Ming Dynasty (1368–1644) poet Yi-Choun[5]:

> *What's the matter with the crickets?* （蟋蟀关何事，）
> *their sad melodies fill the night.* （哀音入夜偏.）
> *So few they are,* （寥寥闲永漏,）
> *yet so loud their song.* (切切乱鸣弦.）
> *It cuts through the breeze and* （乍逐微风断,）
> *coagulates in the drizzle.* （还从疏雨连.）
> *No sleep in sight for the anxious lady within her home.* （闺中有思妇，怪尔不成眠.）

The picture of an tranquil lifestyle emerges vividly out of this painting. The katydids are depicted precisely enough to be easily identified as *Mecopoda elongata* (L. 1758), a widely distributed species in Southern China. Obviously, the ancients observed their behavior, for instance wandering out in the wild during summer (暑则在野) and, by the time chill weather came, showing no hesitation in moving close to human habitation (寒则依人). Crickets and katydids made it possible for people to live with the same sense of poetry as yearned for by Hölderlin.[6]

Numerous poems and paintings relate to the love of insect song in the wild, and many are extremely old. However, the hobby of keeping singing insects in cages for their melody started in the Tang Dynasty (AD 618–906), as recorded in the Chinese book *Affairs of the Period Tian Bao* (《开元天宝遗事》, 王仁裕， AD 742–756):

> *Whenever the autumnal season arrives, the ladies of the palace catch crickets in small golden cages. These with the cricket enclosed in them they place near their pillows, and during the night hearken to the voices of the insects. Thousands of imperial concubines to find the crickets as a good company during their lonesome night.*[7]

It was recorded that in the time of Qing dynasty (1688–1911), during the Yuan Xiao lantern festival in late winter, lanterns were hung throughout the emperor's garden. When they were lit and music commenced, suddenly, cricket song issued forth from the rockery, as crickets placed there were bred from eggs collected the previous autumn.[8] The idea of entertaining the royal family with crickets out of season was extremely clever and demonstrates the delight with which the sound was received.

The custom of keeping crickets is believed to have been imitated by the general populace, gradually becoming a folk custom. In fact, the singing behavior of katydids and crickets was noticed as far back as the beginning of agriculture, when cricket song and behavior became part of the phenology of early farming, used to indicate seasonal changes. This will be discussed further later.

This traditional hobby is still alive today, even in many cities as modern as Shanghai. A middle-aged woman was recently interviewed by one of the authors (Jin) in Shanghai's major Wan Shang insect market, while looking at small singing "Yellow Bell" crickets in cases. She said, touchingly, "My mother is in hospital now, but she still cannot feel comfortable without this lovely song with her. She is used to it and asked me to find one for her".

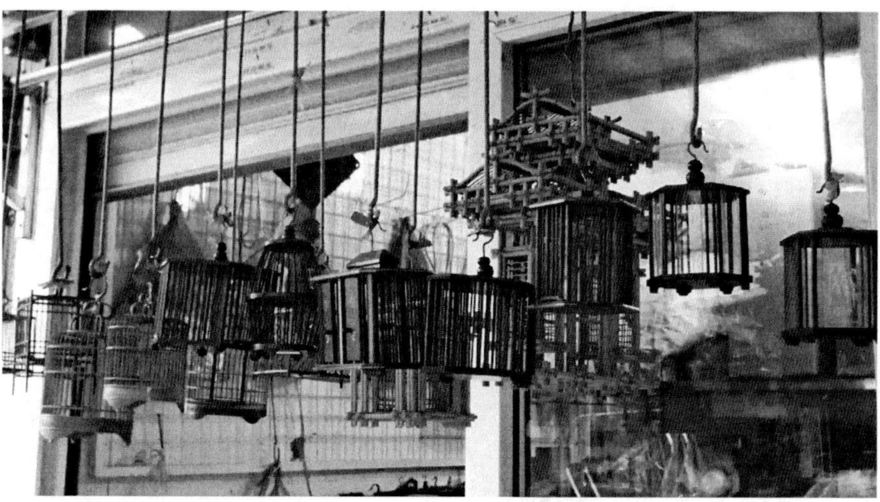

Figure 7.1 A sample of the variety of cricket cages available at the Shanghai insect market. Photo by the Xingbao Jin.

It is quite common in Shanghai at the start of summer for people to hang in their homes small cages made from bamboo containing a singing katydid, called "*jiao guo guo*" (叫蝈蝈), meaning "singing brother" in the Shanghai dialect (Figure 7.1). Many old-fashioned people, especially gentlemen, keep a singing cricket, called "*jin ling*" (金铃) or "*jin zhong*" (金钟) meaning "golden bell", in an exquisitely made case, designed to be carried in a pocket to listen to a song that, as long as they are kept in good condition, is almost continuous. This is especially enjoyed by those who are ill or suffering from insomnia. To some, the cricket song is music from Heaven (天籁). Some taxi drivers even place singing crickets in their cars for their own enjoyment and for the benefit of their customers, especially travelers wanting to relax after a long-haul flight.

Listening to the song of crickets could be a hobby all year-round. Although there is seasonality to their abundance in the wild, there are always at least five or 10 species of singing insects available in the cricket markets. Fans of singing insects believe that the cricket song is helpful calming one's emotions and lowering blood pressure, and some have even used cricket song to induce hypnosis. Although there is no solid scientific evidence for the physiological value of the insect song, the intangible or spiritual value is certain, aligning it with other important cultural phenomena.

Flourishing Insect Market

In Shanghai alone, nearly 20 singing insect markets exist, located inside the flower and pet (birds and fish) markets. Four of these insect markets are the major outlets in Shanghai, spread across the city. During high season for the insects, most of the singing crickets and katydids sold at the market are

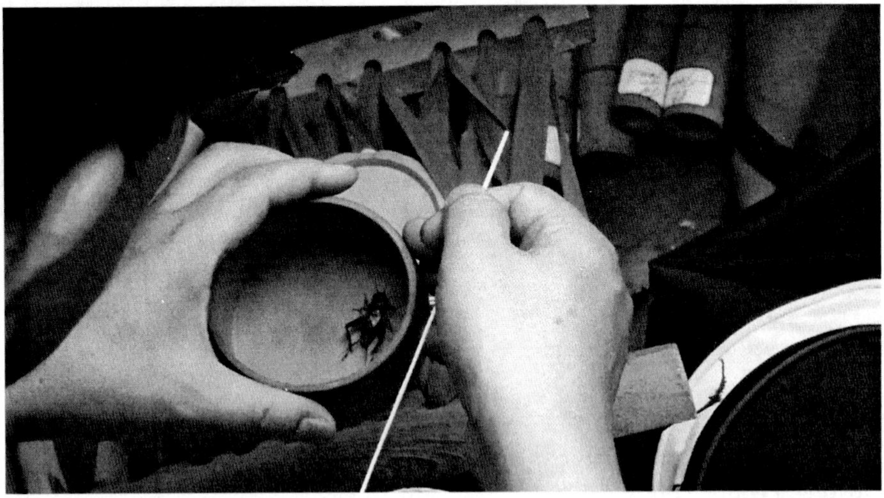

Figure 7.2 Selecting a cricket at the insect market in Shanghai. Photo by Xingbao Jin.

obtained through insect dealers, who purchase the insects from local farmers. Farmers collect the insects from the wild in Anhui, Jiangsu, Shandong and Hebei Provinces. When crickets are unavailable in the wild, those sold at the markets are produced by insect breeders. Breeders are masters that specialize on certain species of crickets or katydids, descendents of ancestral family cricket-breeding businesses, especially in Beijing and nearby Tianjin. Insect breeders concentrate on different species according to whether they operate in the southern or in the northern part of China. Although many breeders depend on the economic income of their business, others are avid enthusiasts, devoting their considerable efforts into breeding crickets as a hobby (Figure 7.2).

According to Shanghai media reports,[9] the annual volume of trade in singing insects in Shanghai is around 200 million Chinese Yuan (ca. 20 million euro), half of which is associated with the paraphernalia for collecting, rearing and keeping crickets, and the instruments or apparatus used in fighting. This pastime has given rise to a unique art form for which many traditional Chinese crafting techniques have been developed.

A Traditional Art Form

Fans of singing insects are often captivated by the great variety of unusual and elaborate utensils associated with the pastime. Utensils can be grouped into two main types: one for singing insects in general and the other specifically designed for fighting crickets.

For singing insects of different katydid and cricket species, there are various containers, which can be divided into five major kinds, distinguished by their external form: cages, cases, tubes, pots and jars with various shapes and patterns for summer and winter, accordingly. Containers can be made from a great variety of material, such as straw, bamboo, sandalwood, gourds, clay, metal, mother-of-pearl, turtle shell, even jade, ivory, silver and gold.[10] Containers can have regional characteristics. For instance, clay pots or jars for keeping both singing and fighting crickets vary in shape and size from northern to southern China, as well as with the skill of the potter. However, no matter how varied in material, shape and size, every kind of container is well designed for both function and aesthetics. Most are handmade. Each container is made with breathing hole, a groove or slot for supplying food and water conveniently, as well as easy cleaning.

To enhance the enjoyment of the effect, several methods have been developed to magnify the sound of a singing katydid. A special resonator is made from bronze wire, used as a cover of a gourd cage. To modify or improve the insect song, a specialist may add a drop of rosin mixture on the vein of the fore wing of an insect, to soften the sound of its song, in a manner used for the "*Er Hu*"(二胡), the Chinese two-stringed violin. This treatment does not hurt the insect, as there are no nerves on the insects' wing veins.

A vast array of utensils have been developed to work with fighting crickets, for feeding, 'tickling,' bathing, mating, weighing before competition, arenas for fighting and even, historically, coffins for cricket champions. The fighting arena is completely different between Taiwan and mainland China, the former is in ellipse shape, the latter is in trough shape, long and narrow.

Controversial Cricket Fighting

In addition to singing katydids and crickets, the majority of fans are keen on fighting crickets, a sport that may predate Emperor Xuan Zong of the Tang dynasty in the eighth century AD.[11] The popularization of the sport, almost to the point of becoming nationalized, can be attributed to Jia Sidao (1213–1275), prime minister during the Southern Song Dynasty. Jia is infamous for merrymaking and engaging in cricket fighting, even when an attacking army had reached the city gates. However, it was this man who authored the first monograph on crickets (《促织经》), which is still the sutra of cricket fighting.

The sport is still popular today in some cities of China. Among the 800,000 singing insect fans in Shanghai, the majority are involved in cricket fighting and the cricket trade. Peak season for collecting fighting crickets from the wild and trading in towns and cities is from mid-August to mid-September, when vendors of fighting crickets at the market increase. Most cricket-fighting enthusiasts are involved in gambling, although purist cricket fans are still insistent on cricket fighting for amusement only. Both cities purchasing crickets, and local towns producing them, benefit economically. In cities such as Ningjing and Ningyang (Shandong Province), the cricket market has become the main tourist attraction and the most important source of local revenue.

An unfortunate aspect of cricket fighting, and the associated gambling, is the criminal cases exposed every year involving huge wagers, principally from big cities like Shanghai and Hong Kong. In some cases it is no exaggeration to say that both human beings and crickets are engaged in the fight.[12] It has been suggested that cricket fighting might be a substitute outlet for natural human aggression[13] and that the fighting cricket is simply another sort of gladiator, just as in boxing.

The ugly face of cricket fighting has brought this traditional sport down to the underworld from its traditional roots, among the aristocrats and the upper classes. At that time, the cricket fighting was looked on rather as an 'art' than an 'amusement'[14] and, as such, can be considered a national sport like bull fighting in Spain and Sumo wrestling in Japan.[15]

There is, however, another outlet for the genuine enthusiasts. Groups of cricket aficionados conduct annual cricket-fighting competitions in cities both in Taiwan and mainland China. These tournaments are highly regulated, having strict procedures to prevent unfair practices. The pairing of crickets is based on a weight system, much the same as human boxing. Even stricter regulations control drugging insects: fighting crickets must be kept inside sealed cases for several hours before a contest.

For many purist devotees, cricket fighting has important philosophical meanings. It is believed that the finest of the competing crickets best display each of the five human qualities or virtues (五德).[16] They are trustworthiness (守信), he will sing when it is the time to sing; courage (立勇), he will not hesitate to fight on meeting an enemy; loyalty (忠诚), he will not surrender even when seriously wounded; knowing shame (知耻), he will not sing when defeated; sensitive to temperature (识时), he will sing differently in cold weather.

Cricket aficionados have anthropomorphized the insects to the extent that they are honored and respected like military generals.[17] At one time in Southern China an especially famous champion cricket was, on its death, placed in a small silver coffin and solemnly buried.[18] The whole process of turning common crickets into warrior generals is even more anthropomorphic. The rich knowledge and techniques based on experience of cricket catching, breeding, caring and fighting is handed down from generation to generation, which has important cultural resonance.

SYMBOL OF NATURE, ROOTS OF CULTURE

According to the British anthropologist Edward Burnett Tylor (1832–1917), the classical definition of culture is a complex whole that includes knowledge, customs, beliefs, arts, morals, law, custom and any other capabilities and habits acquired by man as a member of society.[21] The customs and knowledge associated with singing and fighting crickets are, without doubt, cultural phenomena. The idea of digging out the roots of this culture to serve the conservation of urban nature had been fermenting a long time before we, the authors, began to collaborate on this project.[22]

To many Western people, who pay more attention to large animals, it seems surprising that this unique Chinese cricket culture could come into being. After extensive study in China during the early part of the last century, Berthold Laufer, anthropologist at the Field Museum of Natural History in Chicago, wrote in a special report *Insect-Musicians and Cricket Champions of China*: "The relationship of the Chinese to crickets and other insects presents one of their most striking characteristics and one of the most curious chapters of culture-historical development".[23]

As significant elements of nature, various insects, including katydids and crickets, have captured people's attention since the very beginning of human history. The word "Summer" (夏) and "Autumn" (秋) in the earliest Chinese, as inscriptions on oracle bones (甲骨文) was in a rough shape of cicada and cricket, ca. 16th–11th century BC,[24] reflecting the habit of cicadas to sing during the summer and crickets in autumn. Numerous early drawings relate to singing and fighting crickets, and the tiny, easily caught singing varieties have been indicators for the changes in weather and climate important to annual agricultural cycles. One of the folk names for cricket is "*Chu-zhi*"(促织), meaning "quick-to-weave". People realized that when crickets started singing, winter was not far off, and the woman of the house had to hurry to finish weaving the family's winter clothing. There are many poems about cricket song surprising a lazy wife. One katydid (*Mecopoda elongata*) is even named "*Fang-zhi-niang*"(纺织娘), meaning "weaving lady", describing a song so loud it reminded people of the sound of a working shuttle.

Insects are so small that they are easily overlooked. But they are diverse and widely abundant, and the orthopterans—crickets and katydids—are distributed. They can be seen almost everywhere in villages, towns and cities, wherever land is left unpaved. The well-known fighting cricket *Velarifictorus mikado* can survive even in old houses with falling walls and broken bricks. Their ubiquity is the fundamental reason that cricket fighting as a sport has become so popular in China: its low cost presents no obstacles to becoming involved. The nearly 30 species of katydids and crickets spread across both northern and southern China are the natural source of the singing insect culture. Without them, the tangible natural heritage, this characteristic culture would have nowhere to root.

Although insects and humans are very different life forms, they share the same auditory capacity, albeit achieved using different sound receptors. The sounds produced by male insects are used principally for three types of communication. One is a territorial call to drive off intruders, one is a song for attract females of its own species into its territory, and the third is a courtship song to induce mating. Songs can have other functions too, however. One species of predatory katydid, for example, mimics the sound of other singing insects, such as cicadas, to prey on them when they approach to mate. Insects sing purely for their own survival, rather than for the sake of human enjoyment, despite the proximity of shared living spaces. It is only when these monotonous sounds are received by humans that they are given meaning. The feelings are, in fact, more for the sounds than for the producers, the insects themselves.

Insect song is, to humans, a symbol of nature, indicating the change in season. Most crickets sing at the end of summer and beginning of the autumn. That is the time when the greens are turning to orange, and trees start to defoliate; the cold and long winter is coming. It is a sentimental time for people, especially for those who are far from home. Naturally, cricket song has become a sign to express one's emotion of homesickness, loneliness and reminiscence, especially childhood memories, as many Chinese over 50 enjoy the experience of hearing, catching and keeping them or simply watching others engage with crickets nearby. Various touching poems mentioning cricket song have become classical literature in China.

A well-known poem, for example, "Ode of Mu-lan" (《木兰辞》) by Guo Mao-Qian (郭茂倩, AD 1041–1099), a story about a girl hero named Mu-lan who joined the army and replaced her father. The poem starts with the song of cricket:

Tsiek tsiek again tsiek tsiek, （唧唧复唧唧，）
Mu-lan weaves facing the door. （木兰当户织。）
You don't hear the shuttle's sound, （不闻机杼声，）
But only hear daughter's sighs. （唯闻女叹息。）

Even today, cricket song is the good vehicle to express one's emotion, opinions and communication with each other implicitly and smoothly. The famous poet Luo Fu (洛 夫)[24] from Taiwan wrote the "Cricket's Song" (《蟋蟀之歌》) in 1990, asks the questions "Which cricket is that calling"? (究竟是哪一只在叫？) One can feel the poet's sensibility to his homeland of mainland China vividly in his words. Later, another well-known poet, Liu Sha He (流沙河)[25], from mainland China, replied in his poem "That is this cricket" (《就是那一只蟋蟀》). Most crickets sing during the night when the moonlight is shining. People admire the tiny but vocal tree cricket, saying "if moonlight could be heard, it should sound like this!" as written by the American novelist Nathaniel Hawthorne[26].

Crickets' chirps are so rhythmic that they can be used to mark time, and the rate varies with temperature. Dolbear[27] suggested that the song of the snowy tree cricket *Oecanthus fultoni*, a common U.S. species, was useful as a biothermometer and the relationship became the "Dolbear's law" ever since. If other cricket species were equally well studied, their songs might also prove useful as indicators of local temperature and perhaps even long-term environmental change.

However, insect song does indicate habitat variables in several ways. The songs of ground or brushwood-inhabiting crickets differ markedly from tree-dwelling species. A field survey of common singing insects[28] was carried out in nine localities with differing habitats, distances from the Shanghai central business district, sizes and floristic composition of green space and degrees of human disturbance (frequency of grass cutting, pesticide use, vegetation disturbance and cover of leaf litter).

Results of this study revealed significant information about the management of Shanghai's green spaces. The number of species and abundance of singing insects was positively correlated with natural condition of habitat and the extent of urbanization, especially the amount of vegetation. More singing crickets were associated with more naturally vegetated habitats, and fewer crickets found in more urbanized and less vegetated areas. The number of insect species increased with proximity to the hilly area of Shanghai, and maintenance of urban greens affected the number of singing insect species negatively. More extensive areas had greater insect diversity and locations with less pesticide spraying, vegetation cutting and mowing and removal of dead leaves, improved the chances of crickets surviving because most eggs of crickets and katydids are laid under the soil or dead leaves, inside plant stems or tree trunks.

Some cricket species do occur in recently built urban greens, because the topsoil and plants that were moved into the greens contain cricket eggs. However, they will soon disappear after the application of pesticide and other careless management practices. A few species of singing crickets have been newly recorded in Shanghai since the 2000 survey. The little yellow bell (小黄蛉) *Anaxipha* sp., for example, is found on habitat covered with reeds in a naturally restored pond near Shanghai Science and Technology Museum. This species, which lays its eggs on the veins of reed leaves, is widely distributed across the Orient.

The results of the survey demonstrate that it is not difficult to conserve singing insects in the urban environment as long as a suitable habitat is restored. Although a great number of cricket fans exist in Shanghai, few of them really care about the survival of crickets in the wild, either in urban or rural environments. Despite the fact that most are aware that cricket habitats in urban and suburban areas are vanishing dramatically in the face of accelerating urbanization, few realize the direct effect of human behavior and activity on the crickets and the natural environment. The insect's song could therefore be a good indicator of the 'naturalness' of the urban environment, as nearly all the songs of these common crickets and katydids are quite distinctive.

Most of the refined sounds stridulated by crickets are delicate, like tiny bells. "If a jewel of dew could sing, it would tinkle with such a voice!"[29] This is a sound of silence; only those people who are calm can hear it. It touches one's heart rather than one's ear, like a refreshing drizzle on your face. Many poets call cricket song the sound of home, of childhood, of peace. As human beings, we cannot be satisfied solely by tangible substances like food, clothing and housing. We also have spiritual needs—to enjoy the beauty of nature in various ways, color, form, texture, taste, sound—the 'differentness' of everything. Humans have a need to dwell on earth with the sense of poetry present in Hölderlin's works and to explore the unknown world to find out the wisdom of his surrounding environment, the essential relationship of man to man, and man to nature. This is a psychological requirement, which can, in part, be satisfied by katydids and crickets.

Singing insects are themselves tangible natural heritage, but the way Chinese people appreciate, enjoy and adore them is intangible heritage, relying on the existence of the insects. Research has demonstrated that people respond on a deeper psychological level to just about any natural thing than to most man-made objects, a phenomenon known as "biophilia".[30] Cricket fans looking at the insects more or less anthropomorphically may, in some way, serve to weaken their anthropocentric viewpoints, possibly making them better able to accept new concepts of modern ecological ethics and of the interconnectedness of life. Singing insects may therefore serve as a good vehicle for communicating science to the public.

FLAGSHIP SPECIES FOR URBAN NATURE CONSERVATION

The nature of katydids and crickets has been the basis for a cultural tradition that has survived for more than 1,000 years. Members of the older generation are still able to dwell harmoniously with crickets and katydids both in cities and villages. Six common singing insects share a small courtyard with the author (Jin) in her home in urban Shanghai, made possible by almost a decade of avoiding pesticides and reduced habitat disturbance.

Is it possible to use this traditional culture to encourage more people to appreciate the singing insects and, by extension, to care more about urban nature and even conservation nationally? Currently, many young people cannot even differentiate a cricket from katydid in the wild, far less tell a field cricket song from that of a tree cricket, although they may have had the experience feeding virtual crickets on the Web. Natural science museums have an opportunity and a responsibility to channel young people toward paying more attention to real nature, to understand, to appreciate and to use nature sustainably. It could start with the more commonly encountered creatures, such as crickets, and to solve the gambling problem by transferring interest in fighting behavior to enjoyment of the song. This has been a goal of the Shanghai Science and Technology Museum since 2007.

A nighttime event involving a concert of live insects, called "Listen to the Autumn" (听秋晚会) was held in the museum during the 2007 mid-autumn festival. Hundreds of crickets of different species were hung in cages, cases and on farmyard props, and their singing was accompanied by traditional melody played by a young musician on an ancient Chinese instrument. The music gave participants a feeling of relaxation and peace. One of the audience members, Feng Liao, (Lars Fredriksson) from Sweden, a cricket and Chinese culture lover, was greatly moved. "You feel warm when you hear the song in a place full of snow and ice for more than half year". He had, in fact, introduced this type concert at the Museum of Stockholm prior to 2007 and had attempted to bring home representatives of all the cricket singers from Shanghai market. An Italian attendee was enthusiastic and said "These are the symbols of life and you just feel being safe, when you hear the song of the crickets after a natural disaster". Even teenagers in the audience felt refreshed when they were listening to the insects' continual singing: "The beauty of the Autumn can also be heard too". A citizen wrote a letter to the museum after reading a newspaper report of this concert that "you have realized my dream for singing insects", suggesting that there should be gardens specially reserved for these creatures: a garden without crickets singing would have less vitality.

In late summer and early autumn 2007 and 2008, special night camps were organized by the museum for preschool children to give them an opportunity to search for the insects by listening for their song in field.

The children were excited when they heard the various insect songs and saw their striking forms. They were surprised to realize that such lovely bugs were sharing the urban and suburban spaces with them.

In Taiwan, looking for the ground crickets in the field is still a traditional game for many children. In 2008, the author (Jin) visited a cricket breeding laboratory designed for education in Xinghua primary school in the Taiwanese town of Xinglong. Each pupil was asked to keep a pair of crickets (乌龙仔 *Gryllus bimaculatus*, a common species for sport fighting in Taiwan) from a very young nymph through to adult singer, to experience the entire process of cricket development, understanding a vastly different life form while enjoying the song it produced. A commonly held belief among entomologists, conservationists and museum educators is that both the culture and nature of singing insects need to be spread and passed down to younger generations. Two well-designed exhibitions on this topic were opened in Taiwan Museum in Taipei and Natural Science Museum in Taichung in the summer of 2009. One of the exhibitions of singing insects made by the latter has been traveling both in mainland China and Europe since 2010.

FROM CRICKET CULTURE TO ECO-CIVILIZATION

Nature was once considered merely wilderness and dangerous. However, when nature is used properly, it is the wellspring of culture and allows the formation of civilization. As this example demonstrates, nature and culture are cause and effect. Without natural headwaters, which are the ultimate source of cricket populations, the culture of crickets would have no basis. In China, appreciation of crickets and katydids, whether for their singing or fighting, is a way of observing, discovering and using nature by people with a long history of agricultural practice and rich natural resources. It is an important component of Chinese agricultural civilization. Discovery and recognition of natural beauty, aesthetics and spiritual values have made it possible for these traditions to be passed down from ancient times. It is a good example of the 'creative relationship' between man and nature described by Fernandez-Armesto,[31] through understanding, utilizing and transferring nature to culture through a singing insect.

Exposing the public to the natural source of this particular culture, prompts them to ponder a city rich in biodiversity, meeting their spiritual and poetical needs, is the essence of an eco-civilization. Our natural science museums have opportunity, and a responsibility, to encourage the public to become involved in building eco-cities and helping to create a harmonious society for a better life through the approach of conserving tangible and intangible natural heritage such as singing insects.

ACKNOWLEDGMENTS

The authors would like to give their sincere thanks to good friends and colleagues, Dr Jengtse Yang, , and Dr Zhongxing Yang, , from Taichung, Taiwan, who have shared a passion for this topic with us for a long time, to colleagues from Shanghai Science and Technology Museum, Qiang Si, Manping Liu, Jianhong Xia and Baocung Zhou, who have supported the singing insect in many ways. Great thanks to Hugh Raffles from The New School, an anthropologist with special interests in bugs, who managed to copy the images of the typical Chinese ink painting from the Metropolitan Museum of Art in New York on the author's (Jin) request.

NOTES

1. Richard Register, *Ecocities: Rebuilding Cities in Balance with Nature* (Gabriola Island: New Society Publishers, 2002).
2. Tian-Ji Chen, *Nature and Culture: The Story of Environment, Creativity and Co-Evolution* (Hong Kong: Chinese University Press, 2004). (陈天机，大自然与文化—环境、创造和共同演化的故事， 中文大学出版社).
3. Eraldo M. Costa Neto, "'Cricket singing means rain': semiotic meaning of insects in the district of Pedra Branca, Bahia State, northeastern Brazil", *Anais da Academia Brasileira de Ciencias* 78 (2006): 59–68.
4. The book of Odes, "Shi Jin"(《诗经》: Bin Feng July(豳风.七月), http://www.shigeku.org/xlib/gd/sj305/115.htm
5. The Cricket, "Xi Shuai"(蟋蟀), by Yi-Zhong (颐塚), cited in Zhao-lian Meng, *Chinese Singing Insects and Gourds* (Tianjin Ancient Books Publishing House, 孟昭连，中国鸣虫和葫芦。天津古籍出版社 1993).
6. Friedrich Hölderlin, a German poet who wrote about how man should get along with nature, ("Poetically Man Dwells"). "The Journey" is one example. See *Friedrich Hölderlin: Selected Poems* translated by David Constentine (Bloodaxe Books, Newcastle upon Tyne 1996).
7. Ren-Yu Wang, (王仁裕) n.d. Cited in Berthold Laufer, "Insect-Musicians and cricket champions of China", Field Museum of Natural History Leaflet (*Anthropology*) 22 (1927): 1–15. See also Zao-Lian Meng, *Chinese Singing Insects* (Tianjin: Baihua Literature and Art Publishing House, 2007). (孟昭连，中国鸣虫, 百花文艺出版社).
8. Zao-Lian Meng, *Chinese Singing Insects*. (孟昭连，中国鸣虫, 百花文艺出版社).
9. Jian-Feng Guo, "Little Katydids Sing Out the "Bug's Economy", *Xin Min Evening News* (Shanghai), September 21, 2007, A8 (郭剑峰，小蝈蝈唱响沪上"虫虫经济"。新民晚报，9月21日，A8 版).
10. Xingbao Jin, "Chinese cricket culture: an introduction to cultural entomology in China", *Cultural Entomology Digest* 3 (1994): 9–16.
11. Quan-Xin Zhang, "Historical Tales of Fighting Crickets", *Agricultural Archaeology* 17 (1989). (张泉鑫， 斗蟋蟀史话， 农业考古).
12. Yutaka Suga, See Note 12. *International Journal of Asian Studies* 3, no. 1 (2006): 77–93.
13. Hugh Raffles, *Insectopedia* (New York: Pantheon Books, 2010).
14. Yutaka Suga, see Note 12.

15. Qi-Sen Wang, *Shanghai Cricket Fighting* (Shanghai: Shanghai Literature and Art Publishing House, 2009), (王琪森， 上海打将军 上海文艺出版社).
16. Hugh Raffles, "Cricket fighting," Granta 10, no. 26 (2007):127–144.
17. Yutaka Suga, "Chinese cricket-fighting".
18. Berthold Laufer, "Insect-Musicians and cricket champions of China", Field Museum of Natural History Leaflet (Anthropology) 22 (1927): 1–15.
19. Tian-Ji Chen, Nature and Culture: The Story of Environment, Creativity and Co-Evolution. (陈天机，大自然与文化—环境、创造和共同演化的故事， 中文大学出版社).
20. Xingbao Jin and Alan Yen, "Conservation and the cricket culture in China", Journal of Insect Conservation 2 (1998) 211–216.
21. Berthold Laufer, "Insect-Musicians and cricket champions of China".
22. Yao Zhou, A History of Chinese Entomology (Wugong: Entomotaxonomia, 1980). （中国昆虫学史，昆虫分类学报社). See also Xingbao Jin, "Chinese cricket culture: an introduction to cultural entomology in China".
23. "Ode of Mulan"(木兰辞)，http://poem.8dou.net/html
24. "Cricket's Song"(蟋蟀之歌) , http://www.en84.com/article-5685-2.html
25. That is this cricket (就是那一只蟋蟀), http://www.cnpoet.com/xiandai/n/n0026/n0026_0003.htm
26. Nathaniel Hawthorne 1851. http://entnemdept.ufl.edu/walker/buzz/a00samples.htm
27. Amos Dolbear, "The cricket as a thermometer", American Naturalist 31 (1897): 970–971.
28. Xingbao Jin and Qiang Si, Cricket Community and Population Structure in Shanghai Survey Report (unpublished, 2000).
29. Lafcadio Hearn, Insect-Musicians. Exotics and Retrospectives (Boston: Little, Brown and Company, 1898).
30. Christian Lachel, "Nature Lovers", Museums Journal 104 (2004): 24–27.
31. Felipe Fernandez-Armesto, Civilizations: Culture, Ambition, and the Transformation of Nature (London: Macmillan Publishers, 2000; Chinese version, 2008. Transl. Xue Xun 薛绚译, 文明的力量—人与自然的创意关系. 台北: 左岸文化，585页.

8 Terra Cognita, Down and Under, Living Stones and the Sound of Stones

Reflections on Four Exhibitions

Ulrike Stottrop

INTRODUCTION

Only one year after the United Nations Educational, Scientific and Cultural Organization (UNESCO) *Convention for the Safeguarding of the Intangible Cultural Heritage* was adopted (2003), "Museums and Intangible Heritage" was the theme for the General Conference of the International Council of Museums (ICOM), 2004, in Seoul, Korea. In its program announcement, a new challenge for the museums community was raised:

> *Living in the realm of natural and social environments, people create both tangible and intangible heritage formations that express the continuity of life in all its diversity. It is widely acknowledged that museologists across the world have so far paid great attention to collecting, preserving, researching, exhibiting and exchanging tangible objects, both cultural and natural, in establishing museums as places for research, community development, heritage interpretation and public education.*
>
> *Culture manifests itself not only in tangible forms but also through intangible elements. It is transmitted from generation to generation by means of language, music, theatre, attitudes, gestures, practices, customs and a whole range of other forms of mediation, as well as objects and places in which the ideas of human beings are located. . . . The worldwide museum community now recognizes that it will have to pay significant attention to intangible cultural heritage as well as tangible resources by fostering interdisciplinary approaches. In the preservation of the totality of heritage resources, museums should continue to further their core business of collection, preservation, research, exchange, exhibition and education . . .* [1]

The International Committee for Museums and Collections of Natural History (NATHIST), also focused—next to "Ethics" and "Biodiversity" –on "Intangible Heritage". Gerhard Winter, secretary (1998–2004) and chair (2004–2010) of ICOM NATHIST asked in his letter of invitation the question:

> *We manage nature scientifically in taxa, but all over the world people established their own and maybe quite different "criteria", e.g. according to the medical value or according to the importance for daily life. People make rocks sing and many traditions are centered on animals, plants, rocks and water. How to present these intangible values in natural history museum?*[2]

The contributions related primarily to intangible cultural heritage relative to a natural environment, while others concentrated on intangible elements in a museological context.[3] Hans-Albert Treff of Germany in his speech *"Intangible Heritage: Remarks of a Philistine"* underlined that *"an unburdened term* (Intangible Heritage) *for the ICOM Conference would have been a great challenge to put the "intangible heritage", as created by nature itself, into a debate"*. In fact, he volunteered a first definition of "Intangible Natural Heritage" by distinguishing it from "Intangible Cultural Heritage". Ongoing discussions showed the difficulty in differentiating cultural –from natural heritage.

For the human sciences, "nature" is a conceptual construct and "intangible" means traditions, also in regard to nature. The question may actually be, is the concept of nature a construct of thinking or nature itself? Or even more philosophically, was there no nature before Man? Natural scientists, on the other hand, commonly consider everything as nature except what has been made by man and sometimes even rigorously exclude from nature anything touched by man. Still another possibility would be to consider man as part of nature and human culture as one among many cultures subjected to frequent interference.

Whatever the definition, it will itself be culturally influenced and, as such, may need revising later on, or in another location, relative to the relevant human culture.

For a museum, intangible heritage poses a great challenge. How do we collect and how do we display it? On the other hand, are our objects as such, apart from their materiality, not already carriers of invisible and untouchable connections and meanings? Objects once collected and stored by a museum have lost their original function and context. From the moment they enter a museum they become symbolic tokens that create a relation between the observer and both the untouchable and the invisible, to which the objects point.

This essay looks at four exhibitions in the light of intangible natural (and cultural) heritage. Each one provides examples of a curatorial approach that attempts to go "beyond the object" and demonstrates methods by which intangible aspects might be communicated. Of course these exhibitions, when they were developed, were not specifically intended to cover intangible heritage, neither cultural–nor natural.

BACKGROUND

In particular, with *Terra Cognita* and *Down and Under,* I wanted to explore the standard of exhibiting and collecting in natural history museums. Over the years, I have been interested in, and influenced by, the work of artists who have attempted to go beyond the object in their work, especially those focusing on nature. In addition, we, at the Ruhr Museum—as elsewhere in Germany—over many years in an interdisciplinary way have been discussing questions regarding the presentation of exhibitions and the function of objects. At the Ruhr Museum this discussion involves archaeologists, historians, natural scientists, and museum education officers. Depending on theme, concept and our ideas about media, we collaborate in the design with external architects and scenographers. Aesthetics, style (*Formensprache*) and atmosphere in our exhibitions correspond with the contents and as such we focus on objects as tokens. So the effects of our staging never represent the (supposedly) authentic. We always construct images using the objects.

During the exhibitions *Terra Cognita* and *Down and Under,* I cooperated with Prof Ursula Gillmann,[4] who is deeply involved in questions regarding perception and thus also with matters like the display of the supposedly undisplayable. The team active in the preparation of both exhibitions was deliberately interdisciplinary. Geologists, biologists, ecologists, historians and sociologists all learned not only to combine their different viewpoints, methods and opinions into a single coherent language but also to create a novel concept that allowed for different interpretations.

The original idea for the *Living Stones* exhibition came to me from Prof Peter Pachnicke, who at the time worked as curator at the *Ludwig Galerie, Schloß Oberhausen.* Years of collaboration and many joint projects at the *Ludwig Galerie*, bridging nature and art, brought us together.

The *Sound of Stones* exhibition, finally, was the result of cooperation with Julia Zanke, musician, singer and composer. In 2004, I was invited to present this last exhibition at the ICOM NATHIST meeting under the header Intangible Heritage. Thus opening up the field for initial discussions on the definition of and the criteria for intangible natural heritage as opposed to intangible cultural heritage.

TERRA COGNITA

The geo-scientific part of the Natural Science collection of the Ruhr Museum with in total some 400,000 objects is certainly, not only in quantity but also in quality, one of the largest and best in Germany. To make this wealth of material more accessible to the public was one of the goals of the permanent geological exhibition named *Terra Cognita* (Figure 8.1).[5]

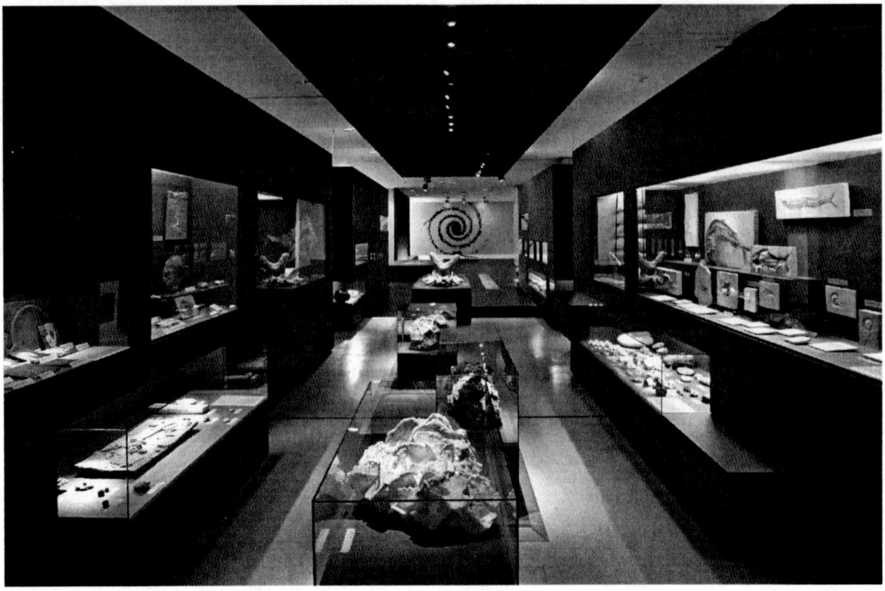

Figure 8.1 *Terra Cognita*: the central axis is 40 m long and leads to the theme rooms of the exhibition. The ceiling-high showcases are filled with fossils from famous paleontological sites. One end of the exhibition space features the spiral nebula of our galaxy and the "*Wunderkammer*" of creation—the initial point of our travel through nature and culture. Photo courtesy of Michael Rasche, Dortmund.

The objects themselves in their beauty, fascination and mysteriousness were at the center of the exhibition. The aim was to show them with their various interpretations and readings and thus to open up new points of entry into the natural heritage. The exhibition was concerned with natural collections in their threefold relevance: proofs of scientific understanding, sources of artistic inspiration and museological documentation. The strategies of display typical of cabinets of curiosities and "*Wunderkammern*" were used and developed. Lyrical and prose quotations accompanied the presentation and offered in their peculiarly alternative entry points for observation and experience. The rustle of the wind and the roar of water, the zing of wing-beaten air, the noise made by sundried sand trickling down a rocky slope, the sound of love, or, a little more prosaically expressed, reproductive noises—recorded sound installations like these provided additional sensory levels.

An integral part of the exhibition concept was an ongoing accompanying program in different media. *Interventus Australis* was an exhibition series of contemporary Australian art within *Terra Cognita*.[6] The artists were free in their choice of media and in the form in which they planned to intervene with the collection. They were invited to an interdisciplinary working process with the scientists of the museum and the art curator, Rafael von Uslar.

Thereby we perpetuated the tradition of the historical concept of a cabinet of curiosity, which combines natural history and the various arts.

"Klangführungen" (sound-tours) were led by a musician who performed a series of set pieces, each of which was a musical response in sound to different sections of the exhibition, including instrumental music, body percussion, humming, singing and recitation. In the history of music, nature and creation were always sources of inspiration and so other musical tours working with choral and Baroque genres were conducted as part of the broader program.

Events like these, the permanent sound installations in the exhibition, the highly poetical quotes, the striking beauty of the objects in their setting moved and touched the visitor. Since the objects did not, as so often in exhibitions, play only an accessory role, they pointed to the fact that what makes them into museum pieces, is that they represent the unseen.

Over and over again artists and authors, architects and musicians have been inspired by nature and searched the proximity of the natural sciences. To them, natural objects in their aesthetic dimension and beauty are subjects of contemplation and sources of inspiration. They move something deep inside. Art, music, language and literature thus become tools to express of the intangible. As evolved forms of creativity, they constitute the basis for cultural evolution. The aesthetics of the objects was the prime criterion for the objects' selection in this part of the exhibition: their materiality, their fragility, their rhythm, their ornamental presence. Views of the graphics of the biologist Ernst Haeckel,[7] of the multitudinous shapes of petrified sponges or perfectly shaped crystals, allowed visitors to enter into fascinating transitions between nature and art that range from the Book of Revelation of John, describing the heavenly Jerusalem built on 12 gems, via medieval gem mysticism, to the surrealists of the 20th century.

Showcases with themed drawers showed the manifold aspects of the intangible heritage of human culture in the museum context. Letters, for example, record a passion for research and collecting. Herbarium leaves document the wanderlust of nature lovers from the educated middle class, particularly in the 19th century. Painstakingly detailed and in miniature script, labels for butterfly wings refer to a need to organize dictated by the human intellect. Many of the containers compete visually with their contents, as the matchboxes, containing snails, from the 1920s and 1930s demonstrate.

A medieval image speaks of dwarves on the shoulders of giants reminding us that we never start from scratch but rather build on traditions of perception and thought. A library with tables and slides, developed with the historian Stefan Siemer, seized this concept. Conceived as voyage through the past of present-day science, it encouraged visitors to explore a variety of worldviews and their related observations of nature.

Terra Cognita didn't exclude points of view that are much older than the current scientific ones. For the ancient Germans, the event that brought the belemnite fossil into being was the God Thor hurling a thunderbolt. Science tells us that belemnites are the petrified internal skeletons of extinct squid.

There was also room for fantasy —the clam *Trigonia costata*, for example, called the "Motherstone", resembles the female vulva and was believed to stimulate fertility.

So far, we have only considered the intangible cultural aspects as applied to natural objects. By contrast, natural objects, formed by the intangible workings of nature such as water and wind, heat and cold and, especially, time represent an intangible natural value, as do particular life events and instants in a life, when petrified or inscribed in stone.

Focusing on the natural science view of the collection, we presented a selection that linked in its multitude to the incredible dynamics of the planet Earth: lava rocks, petrified lightning, deposits transformed into folds, a piece of sea soil shaped by waves or the imprints of a 270-million-year-old rain shower, transformed this interplay of change, translocation, superimposition and the effects of endogenous and exogenous forces into an immediately visible spectacle.

Our knowledge regarding the history of life stems from research of the remains, the fossils. The only objects presented were those that, in their entirety, depict both the design and the diversity with which nature 'organizes' life.

And what is now presented in a petrified form, once was highly vivid: running, feeding, digesting, biting, and breeding to name merely a few of its very natural expressions of life. Objects like the ammonite impressed by a Mosasaur, or the weighty excrement of a dinosaur tell these tales. The fact that an individual needs to feed in order to survive and breed was conveyed audiovisually: the soundbites of copulating and feeding animals and humans together with two short movies, one featuring sleeping animal offspring and the other showing the path from lamb to roast dinner.

An article from the regional newspaper *Saarbrücker Zeitung* entitled "Exemplary Museum in Essen" in 2001 reviewed this innovative concept with the words:

> *The novel permanent natural history exhibition* Terra Cognita *which, nourished by one of Germany's best and most extensive geo-scientific collections, impresses the public in a breathtaking way: no parade of ammonites dominated by some distracting time-line, but rather a confident theme-park with diverse interpretations and readings. That is the Art of exemplifying.*[8]

DOWN AND UNDER—"NATURE–CULTURE" OF THE RUHR AREA

Those who enter the Ruhr Area by its northern highways get a clear impression of—in a geological sense—the newest mountains: the waste rock piles. Like landmarks, they dominate the regional woods. Depending on the perspective, they might even look like mountain ranges. Nowadays they are green areas of local recreation, were in summer and wintertime, lots of strollers meet and many leisure time activities are on offer.

Excrescences from mining as well as from the iron and steel industries are meters thick and cover large areas of natural ground. The geological maps of the Ruhr Area characterize these man-made sediments as the latest deposits in geology. They are ashes, cinders, waste, debris and slag—technologically generated substrate. And yet, precisely those large-sized former mining and industrial areas form a basis for new life: an artificial biotope for plants and animal, whose natural biotopes have disappeared or a new biotope for immigrant plants. So-called "industriophytes" reached the area with the advent of the global transport of raw materials. Mountains of cinders produced by the iron and steel industry turn a sunny yellow, when the *sedum acre* is in bloom and glowing coke is replaced by red flowering *Epilobium augustfolium* and *Geranium robertianum.* Flora and fauna have found a special niche in the Ruhr Area. The toad *Bufo calamita,* for example, is nowadays part of the indigenous fauna in a former smelter. The Ruhr Area was by no means a uniform biogeographical region. Nevertheless industrialization has transformed the region into an 'ethno-ecological' unit.

Nowadays the appearance of the Ruhr Area can be characterized as an industrial landscape marked by extensive human interference. At the same time it contains elements of a post-industrial natural landscape. Such a complex structure demands new innovative concepts, which are difficult to harmonize with conventional scientific forms of observation and representation. So the study of the contemporary Ruhr Area landscape is inevitably a study of man's relationship with nature. The notion of 'nature-culture,' meaning nature influenced by culture, reflects the interference with nature and also the perception of it.

Certainly all these aspects could be explained in an exhibition by images and commentary. But if these aspects are to be reflected and explained by the collection itself, one has to think about new criteria for collecting and representing. These must break with the normal conventions and categories of natural history collections.

For our exhibition *Down and Under: Nature-Culture of the Ruhr Area,*[9] (Figure 8.2) we examined different aspects and elements of intangibility, for example the memories of the inhabitants of the region regarding the 'smell' of the Ruhr Area. In a call for contributions in the local press we asked readers: What does the Ruhr Area smell like? What kinds of memories emerge when you smell your way through factory halls, kitchens and living rooms, parks and gardens? We were looking for people who treasured particular smells from the past—be they everyday ones, like the odor of an industrial area or the work clothing of the blue collar workers, the odor of smoldering coal pits and mild summer evenings; the aroma of local food and drinks, the smell of working days and holidays, the scent of the Ruhr Area.

A surprising result of our call for contributions was that the majority of the memories were of pleasant fragrances, not overlaid by recollections of the smells usually connected with the Ruhr Area. The more unpleasant, stinking smells proved more difficult to describe. So we collected and presented samples of those smells: the work clothing of aluminum workers, miners, brewers, and employees in sewage works. And we asked our colleagues from the department of chemical investigation to produce artificial samples of the typical bad smells of the old Ruhr Area.

I believe that the sound of a landscape is an essential, if intangible, part of its identity. What does the Ruhr Area sound like? I heard of Siegfried Saerberg for the first time on a radio program. He cannot see but hears all the better. I contacted him, and we both agreed that the attentive hearing of a blind person would be a valuable contribution to the exhibition. He was on the road for more than a year, using his professional equipment and this white stick. He described his experiences in his contribution to the catalogue as follows:

> *A soundtrack of almost a hundred hours was produced in this way: the sounds of nature and landscape, of individual people or animals living in the Ruhr Area and also interviews. In every single sound recorded in the Ruhr Area a droning, humming and the tinkle of civilization is mixed in: the engine noises of cars, airplanes, trains and ships, the voices of strolling pedestrians, cyclists with portable radios on their handlebars, who bring even more civilization to the landscape, the sounds of human labor, machines which blow, hammer, saw, cut. Last but not least the church bells at noon and in the evening which still remind us of the idyllic villages of former times. Among all these sounds are mixed in: the twittering of birds in springtime, the chirping of crickets in summer, the rustle of leaves in autumn and the cracking of branches in winter. Silence is rare in the Ruhr Area. There is a permanent human interference with nature, but also human care for nature: toads are saved from being run over; hedgehogs are rescued from the winter cold. Maybe the more people love nature, the more it is threatened or no longer intact?* [10]

To present the landscape, we initiated a project in cooperation with 48 photographers. We drew a 10 by 10 kilometer grid over a map of the Ruhr Area—with the Ruhrland Museum[11] in the center. Photographs were taken at the 48 points of intersection. Our aim was a visual inventory of the Ruhr Area. This inventory was not centered around a perspective of special topics, nor around a certain interpretation of the landscape. All photographs were taken at the same time. Following an item is a local newspaper, we found the 48 photographers and on November 21, 1999, at 3:30 p.m., they took shots of the whole Ruhr Area—the earth and the sky and, measured with a compass, views to the north, east, south and west. We had 288 snapshots, which document the Ruhr Area as an objective 'patchwork' of towns, landscapes, industrial areas, fields, woods, streets, railways, waterways. It was surprising for us to see how clearly the old, pre-industrial cultural landscape and the different morphology of the current landscape became visible. Another unexpected benefit from this approach was that an intangible feature like temperature was made plainly apparent. On November 21, the hilly countryside of the southern Ruhr Area was covered in snow, whereas the northern part showed the golden light of a sunny autumn day. The industrial area was visible only in one sole photo, but in nearly all the rest there was green, shimmering 'nature-culture.'

We collected and examined samples of the soils of the Ruhr Area, made from the remains of industrialization and with traces of technology and human use. In cooperation with a specialist in anthropogenic soil, we collected samples of manmade industrial culture: ashes, cinders, coal dust and waste from foundries and construction. A cross-section was taken from the terrain of a former iron works. Cinders from iron works, construction waste, rubbish from a coking plant, scrap metal and other production debris left traces here.

Preserving jars, filled with soil samples polluted by heavy metal, had been collected from the terrain of a former zinc smelter. The municipality took possession of the urban wasteland in the 1980s and had already started a housing project for young families when the heavy metal pollution damage became known to the public. The costs of restoration amounted to 58 million Deutsch Marks—to be paid by the local authorities. A geologist will see those samples as anthropogenic sediments, a social historian or a cultural anthropologist will see them as an example of a failed vision.

Other new collections could be characterized by the title "culture meets nature". We made 15, 60-cm-square plaster casts called "Underground Life": showing access to a gas conduit near a molehill or "In the Snow": traces of horses, dogs, cars and men at the edge of a field in the southern part of the Ruhr Area, to mention only two examples.

We collected water samples to demonstrate the hydrolic system of the Ruhr Area. They included rivers, lakes, canals, waste water, pumped water and industrial, drinking, mineral and polluted water. This collection explained in a simple way the complexity and also the problematic nature of the local water system. Most visitors were surprised by the many different kinds of water existing side by side and were stunned by the colors: green, yellow, orange, brown—murky or clear—the shape of the containers was associated with different kinds of nature. With the use of aesthetics, the perception of dirt and cleanliness and powerlessness in the face of contemporary pollution can be made a subject of discussion.

The ingredients of the air in the Ruhr Area, including ashes, dust, substances from mining and industry had been measured and documented by the regional environment agency. We put them all together in one list. The hundreds of chemical substances which until the 1970s were emitted into the so-called 'endless sea of the air' made a frightening comparison with the air's natural composition.

Diseased lungs in formaldehyde and thin sections of lungs, as well as X-rays showing miners' silicosis, are rather unusual objects in a geological exhibition. Still, a geologist would categorize the silica in the lungs of miners as "Aeolian sediments". But from another point of view, these objects stand for illness and death.

Not much is known about nature in the Ruhr Area before industrialization. Industrialization came with such brute force and ignorance to this region that in consequence hardly anyone thought of saving samples of a state of nature that might be irretrievably lost. Only a few people with an interest in biology examined the region in a scientific way. There are depictions and narratives by poets and travelers, some old pictures, photos and maps—they allow us to form an opinion. There are historical documents and collections, which also help us. With our contemporary knowledge of ecology, we can reconstruct many aspects of nature. In addition, employees of the regional environment agency documented the effects of industrial emissions on the plants of the Ruhr Area in photographs at a time when dust and soot were still a part of everyday life. For example, photos from 1963 show daisies and peaches from the Ruhr Area in comparison with plants of the same species found elsewhere. The ones from the Ruhr Area are smaller and covered with carbon black.

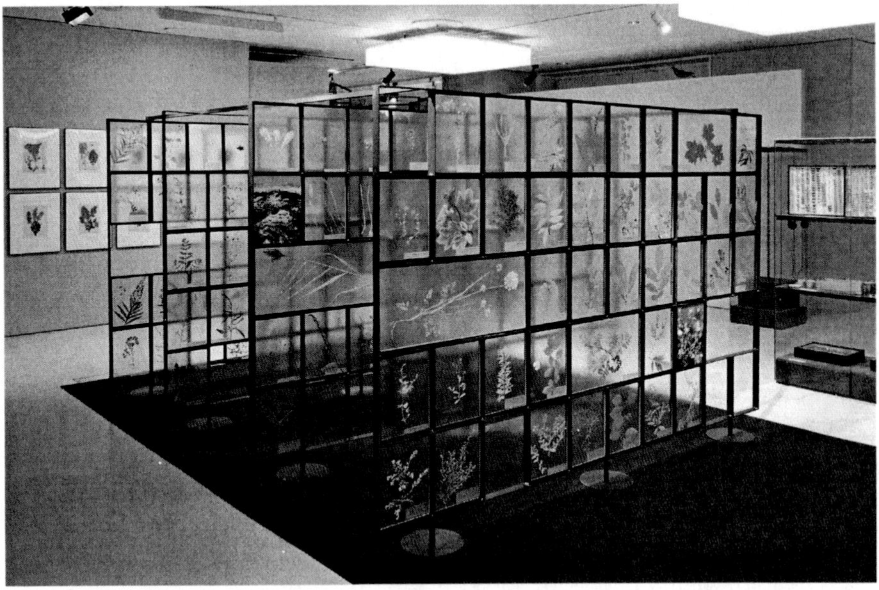

Figure 8.2 Down and Under: a sample view into the flora of the Ruhr area—new communities of plants growing on industrial wasteland. Photo courtesy of Tom Bisig, Basel.

We documented today's flora of the Ruhr Area with classical herbaria. However, of prime importance to us was not their biological systematics but rather their cultural valorization relative to their migration background. Nature of the Ruhr Area is directly or indirectly the result of human interference. A special perception of nature and its value is always attached to a certain significance, which varies with aesthetic perception, cultural needs and social necessities. In each different type of culturally created forest we collected a leaf from every tree within an area of 50 by 50 meters. The number of leaves ranged from 1,645 birch-leaves from a forest that grew spontaneously on industrial wasteland, to seven exotic species of tree in the Park, created by the famous industrialist Krupp.

When coal and steel plants were closed for economic reasons, the disused industrial habitats were left open to be re-conquered by vegetation. Thanks to their characteristic qualities, developed in other places and in other geological eras, entirely new communities of plants grow on industrial wasteland in close proximity to one another. These plants were collected in another herbarium. The rich variety and composition of the so-called 'industrial nature' in the Ruhr Area is as unique as many other natural habitats on earth, although it has only existed for a few decades. Industrial flora consists of archetypes, native species that settled in the region in the wake of the last Ice Age, new plant species that arrived after 1500, and industrial species that came here after the onset of industrialization.

Plants by the roadside are compelled into a particular pattern of movement by the strong drafts produced by cars passing by at high speed. One of the things we wanted to do in the exhibition was to capture this movement in the living plant by closing a flower press on to it at the precise moment when it took on this 'unnatural' shape. The result was a kind of a 'snapshot,' an instantaneous documentation. Incidentally, as a document presented as a 'picture,' the roadside plants in frames can be seen as being in the tradition of studies of nature like the famous painting "The Large Turf" (1503) by Albrecht Dürer in the permanent collection of the Albertina in Vienna.

With the plants living on the edge of the roads, we tried to visualize a physical phenomenon: kinetic energy. In another installation we tried to capture time. In front of the museum, there were some Amelanchier, small trees, green in summer, which belong to Family Rosaceae. The word 'amelanche' comes from Gaelic and means small apple. Over the course of a year, we collected and pressed a twig every Monday. Later, we displayed them individually in glass plates in a circle hanging from the ceiling. With this method we visualized both, the seasonal plant life as well as the explosive force of life as expressed in buds, flowers, and fruits unfolding in the course of weeks.

According to Ursula Gillmann in the exhibition catalogue:

> *The architecture of the exhibition aims to visualize the relations between objects, space, and visitors and to translate the conceptual themes in spatial constellations . . . The architecture adapts to the characteristics*

> *of the Ruhr Area, the patchwork of a seamless vicinity of different scenic elements, the fragmentation and disintegration by traffic ways. The design plays with paths and movements of the visitors in space to enable different viewpoints and perspectives, to allow surprises and to animate discoveries in order to incite, not only a reflective, but also a playful and associative contact with the exhibition.*[12]

To conclude: at the Ruhr Museum, we focus on nature in its 'being there,' in its concrete form of existence, in its everyday reality, into which mankind is unavoidably interwoven. A characteristic feature of all the collections is that they are not only documents of various aspects of an ecological system, but that they also display a visual and aural aesthetic that includes documents regarding the olfactory nerve. Especially for a museum in the Ruhr Area, it makes sense to perceive man and his activities as a part of nature instead of a contradiction. Nature is at once the origin, precondition and outcome of human existence. To observe nature, to experience, describe and analyze it with scientific methods and to dissect it, is a practice formed more by the cultural continuum of the observer than by the subject of observation itself. Being aware of this will open up new strategies of collection as well as presentation, which will outstrip the hitherto well-known types of natural science exhibition.

LIVING STONES—NATURE AS ARTIST

Living Stones—Nature as Artist[13] was on exhibition at Ludwig Galerie *Schloß Oberhausen*, an art museum in 2007 and in 2009 at the new Ruhr Museum. Its purpose was to communicate the wonderful art of nature. The exhibited stones span 2 billion years, and the photographs of Albert Renger-Patzsch, Alfred Ehrhardt, Karl Blossfeldt and Thomas Wolf show that both nature and art embody the creative force (Figure 8.3).

The photographic composition *Gestein* (Stones)[14] by Albert Renger-Patzsch (1897–1966) shows the enormous and endless cycle of growth and decay. He found samples of fascinating formations from all over Europe, embodying and displaying the dynamics of internal and external earthly forces. He was particularly inspired by the eroded coastline of Brittany, chalk cliffs in Normandy, Norwegian glaciers and fiords, the Swiss Alps and Italian marble quarries. To the less geologically trained eye, his images often appear abstract. Renger attempts to assemble a landscape through the interconnection of his images and to meld the various aspects into a coherent experience. He included scenic overviews to allow contextual location. Already in his early work, he had effectively used cut-outs to extract details like lines, planes and the abstract or ornamental beauty of material textures. He, in a way, enters the objects to discover their Being. [15] What Ernst Kalai wrote in 1929 equally applies to Renger's later work:

> *The Renger photos communicate Being without the distraction of incidental charm and surpass merely interesting coverage. They are filled with a persistent empathic searching love and a reverence only found among philosophers, persons with a wealth of life experience and gifted artists. The ethos of the world-view in Renger's photos is as sincere and valuable as in old master paintings. Their imaginative completeness and beauty results from both profound vision and technical perfection.*[16]

Karl Blossfeldt[17] (1865–1932), with his famous photography of plants and Albert Renger-Patzsch defined the so-called "new objectivity" *(Neue Sachlichkeit)* as photography of organic shapes.

Alfred Ehrhardt (1901–1984) was an organist, choirmaster, composer, painter and art teacher before he turned to photography and film. In his photographic work, he followed in Renger's footsteps but added a transcendent dimension to Renger's materially perfect shooting technique. He attempted to capture and display the 'inner life-force', that he, like Renger, saw in the multitude of shapes and beauty of both organic and inorganic nature. His central concern was the spiritualization of matter. Ehrhardt concentrated on inorganic matter and studied the changes resulting from the combined forces of wind and water in tidal sediment structures. He also devoted his attention to innumerable shapes of Icelandic volcanic structures, the calcified remains of mussels, snails and corals, and the geometry of crystals. He too used cut-outs but used images of organic wholes. The dynamic elements of his subjects, such as wind and water, were recreated by his experiments with light and camera angles. Crystals, especially, were portrayed as a unity of light and form.[18]

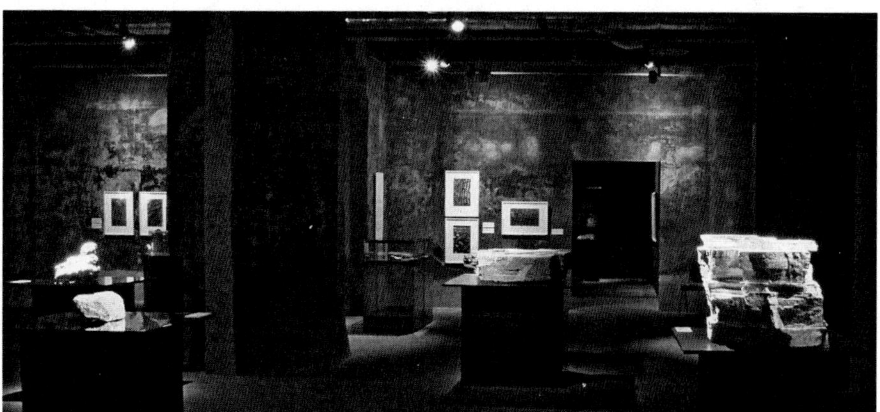

Figure 8.3 *Living Stones*: exhibition in the dark space of a former coal bunker. Photo courtesy of Rainer Rothenberg, Ruhr Museum.

As curators of the exhibition, we used the ways in which the processes of geological historical cycles were shown in Renger's work, as a blueprint for the exhibition. We chose 70 stones from our geological collection to support the idea of nature's own creativity. In the dark space of a former coal bunker, each one was displayed separately, placed on a mirror and spotlit, to emphasize their sculptural qualities. As the beauty of the stones is revealed, they start to tell their story. Visitors to the exhibition, excited by the beauty of the objects, were given the opportunity to explore each stone's unique and extraordinary biography. The visitor could read a story which began "on an autumn day 270 million years ago. . . . " This was the story of a stone exquisitely covered in the fossilized leaves of *Glossopteris*.

On two large screens contrasting films were shown: on one side, the enormous energy of erupting volcanoes conflicted with, or counterbalanced, the other—the still beauty of eroded scenery from Francis Ford Coppola's film *Koyaanisquatsi*. Together with the supporting music of Philip Glass, they developed a suggestive force that captivated the visitor.

Thomas Wolf (born 1967), a German photographer, took pictures of each single stone, while being guided by its very own biography. The digital images and cut-outs were pasted into a film, minimal music added and shown in a coal bunker. By switching vantage point and zoom, Wolf created almost poetic aesthetics. In his images the stones showed, in their materiality, the forces that created them.

The *Living Stones* exhibition ended with a cultural history epilogue: a medieval lead crystal cross and a huge rock crystal pedestal juxtaposed against ancient Chinese scholars' rocks and bizarre rock formations, facing one another. With its lucent brightness, the invaluable cross refers to the triumph and reign of God's son and represents a view of the world in which man dominates nature.

> *And the dread of you shall be upon every beast of the earth and upon every fowl of the air, upon all that moveth upon the earth and all the fishes of the sea; into your hand they are delivered.*[19]

Still, over time, different opinions were heard from within Christianity that searched for God within nature. Theologians like Albertus Magnus and Hildegard von Bingen are merely examples. They influenced thinkers like Herder, Goethe and von Humboldt. God is not revered as a person but adored in his creation in which He reveals Himself. By this, observation of nature becomes a sensory worship.

> *My Teacher was my Mind, the Teacher of my Mind was my Eye and the Teacher of my Eye was the Mountain.*[20]

The highly educated civil servants and scholars in ancient China loved to collect bizarrely shaped, naturally enchanting stones and displayed them on artfully crafted wooden pedestals. They described them in poems and essays. They considered the dense limestone the most valuable. When hit, it produces a bell-like sound.[21] The sympathy for stones, though—common in ancient China—was much more than a mere fascination for the multitude of colors, patterns, shapes and structures found in the various stones and rocks; they were stamped with centuries-old cosmological and religious concepts, as well as man's attitude to nature.

Over 2,000 years ago, the Chinese had already started to arrange peculiar stones in cosmologically engineered palace gardens. As monuments in a garden or on pedestals in scholars' studios, the stones and rocks displayed in this way were miniature symbols of well-loved mountains and rock formations and, as such, ultimately of the entire world of their contemplative life. For the stone-adoring Chinese, it was an enchanting experience to submerge into their convolutions, branches and cavities. Such stones display the recognition that all matter, alive or not, is being percolated with a global vigor.[22] The scholars' rocks symbolize this view of nature.

THE SOUND OF STONES—AN INTERACTIVE EXHIBITION

Finally, I want to describe an interdisciplinary exhibition project between a musician and a geologist. It was planned for Essen's Mineral Museum, which is part of the geological department of the Ruhr Museum. Our plan was to offer our visitors the opportunity to create their own experience with the special sound character of stones and to tell them about the history of using rocks as musical instruments, as well as the meaning and importance of sounding stones in cultural history.

The use of resonant rocks seems to be as old as human history. In 1997 the Pitt Rivers Museum of the University of Oxford published a compendium entitled *Rock Music*.[23] The publication covers such apparently disparate fields as music, prehistory, ethnography and petrography. Rock gongs are found worldwide and are known to have been used since Upper Paleolithic times. They are natural stones and are not artificially tuned.

> [They] *resonate when struck and show evidence of human use as so called idiophones. They seem to be of any shape or size. Several different notes can usually be obtained from a single boulder and resulting from percussion an abraded area may show on the edge of a rock; alternatively hollows or "cupmarks" are formed.*[24]

The compendium cites rock gongs from all over Africa, from Canada, U.S., Brazil, Colombia, India, Thailand, Vietnam, Australia, France, Guernsey, Ireland, Italy, Portugal, Scotland, Spain, Sweden, Wales and Oceania. A major problem in putting rock gongs into their social context is that of dating. Many rock gongs are found close to paintings and engravings. But only a few remain in use at the present time. They are used, for example, in signaling, because the sound carries a considerable distance. Some are known to be used, or to have been used, in initiation and circumcision rites or in farming ceremonies. Some resonant boulders are connected with rain-making and others are played to accompany singing and dancing, because they are excellent talking drums.

Beside the gongs we have so-called 'lithophones,' artificial musical instruments made of stone. These are a series of resonant, tuned stones, suspended horizontally or vertically and chromatically arranged. In the cultural area of East Asia chime stones were common and well known.[25]

> *Rock gong . . . were integral elements in certain religious rituals and celebrations . . . It is the attribute of a "voice" which associates mystery with such rocks: the voice of the rock is believed to be the voice of an ancestor or other spirit with power to summon the supernatural, and as long as it is believed that inorganic nature is not dead, but may be the vehicle of power –the rocks may be treated with awe, even veneration.*[26]

Konei, the Emperor Yao´s court musician, from 2300 BC describes this in a chant:

> *When I smite my musical stone—be it gently or strong,*
> *Then do the fiercest hearts leap for joy, and the chiefs do agree among themselves*
> *When you make to resound the stone melodious,*
> *When ye touch the lyre that is called Ch´in,*
> *Then do the ghosts of the ancestors come to hear.* [27]

The oldest complete lithophone in Europe was made in 1785 in the English Lake District. It can be seen in the Keswick Museum and Art Gallery, and it seems that the inventor, Peter Crosthwaite, had seen or heard chime stones when he worked for the East India Company and captained a frigate operating the Far East. He used local slates.[28]

More elaborate ones, known as rock harmonicas, were made later in the same area. The most famous is the rock harmonica from Richardson and Sons with five chromatic octaves. After 13 years of construction, it was finished in 1840 and was played at two concerts in honor of Queen Victoria.[29] Lithophones with horizontal stones are still built in Europe today.

This historical background was displayed in a room with wall texts and pictures. We also showed a self-made *pien ching* and a very famous modern chromatic lithophone from Germany, which was conceived and produced for the stage performance of *Antigonae* by Carl Orff. It is made from Solnhofen limestone.

Our planet is full of the sounds that nature creates with its own stone instruments. Sounds, which form the geographical keynote of a landscape: the sound of stones in the waves pounding on the seashore, the rolling and rumbling of pebbles in a river, the howling wind blowing over the crest of mountains. The world is sound: with every grain of sand trickling to the ground, with every glacier shifting forward and with every rumbling and gargling bluster of volcanic activity. For the exhibition, we created objects to imitate these natural sounds of nature. A rocker was filled with fine grained gravel, a big funnel hung from the ceiling, filled with sand that trickled out; a wine cask, filled with boulders, as well as pieces like a stone comb, on which to blow (Figure 8.4).

Other materials for this experimental segment of the exhibition were basalt, limestone, granite, roof slate, travertine, marble—and thousands of larger and smaller pebbles. All the stones we showed were ordinary slabs and flagstones (or pieces of them) as used in the stone industry. They could be played with drumsticks or smaller stones, or—the pebbles—with the feet, stomping in rhythm. Baskets were filled with various plates of travertine and limestone to make one's own lithophones. The rough surface of a kind of plutonic rock invited the visitor to rasp with other stony material or sandpaper with different grain size and a chain with pumice stones created a very soft and tender sound when set in motion.

Astonished by all the different notes they could hear, many visitors compared the stones and asked the reason why some rocks resonated when struck. And all of a sudden we were in a living dialogue about the material: Is it the physical texture of the rock as well as its shape and size? What are the components of the rocks? Is the composition relevant or irrelevant? What is the role of the grain size?[30]

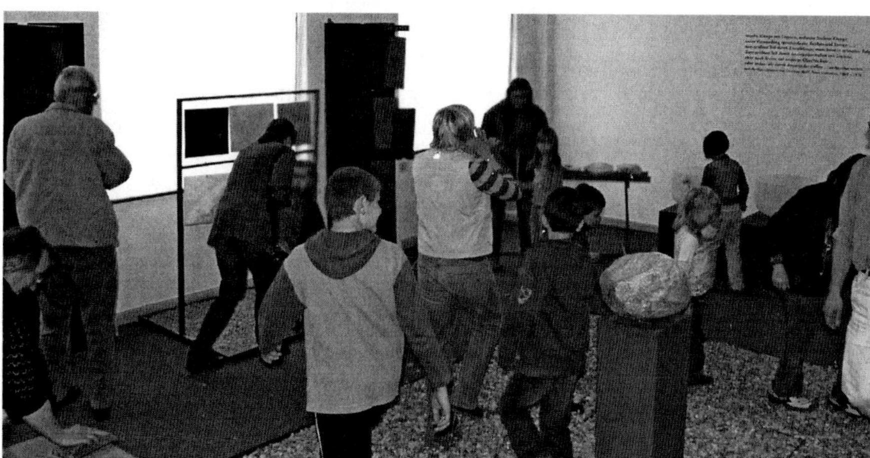

Figure 8.4 Sound of Stones: hands and feet on. Photo courtesy of Rainer Rothenberg, Ruhr Museum.

Another highlight of the exhibition was a sculpture made by Elmar Daucher, an artist, who first produced sculptures to be played percussively. One day after a rain shower, he walked through his outdoor studio, and, wiping the rain off a sculpture, he discovered, that he could make the stone "sing" just by running his hand over it and causing it to vibrate. Hearing this "spherical" sound, basic instincts seem to be touched.

Modern compositions by Stefan Micus with and for the sound-producing sculptures from Elmar Daucher were part of a collection of "stone music" to which our visitors could listen in a special "meditation room". For example, there was ethno-music from Rapa Nui/Easter Island or electromagnetic vibrations (changed so that we could hear them) from a pulsar in Cassiopeia A and from Miranda, the inner moon of Uranus. Most of our visitors took one of the pillows we offered to relax on the carpet. In a bed of gravel, we presented well-rounded boulders from the Baltic Sea; a plate with ripple marks; a ventifact and a stone scratched by glacier—the results of geological processes that had produced sounds at the moment they occurred.

Many composers use stones in their work. Siegfried Saerberg and Thomas Zwerina, two blind musicians composed a modern score for the exhibition, called *Stone Rag—La Partitura Sulle Pietre,* a composition for three or five bricks and piano in C minor. They performed it at special events.

Normally a geologist can analyze minerals and rocks, but they can't render the objects' spirit tangible. This exhibition demonstrated how an interdisciplinary approach can make intangible heritage come to life. The Asian cultural approach to this topic, especially, helps us widen our horizon about the nature of rocks and stones.[31]

FINAL OBSERVATIONS

The descriptions of the four exhibitions provide an indication of how immaterial values may be communicated and can be documented in new collections. Dependent on readability and point of view, natural objects carry certain invisible and untouchable elements. Such elements may relate to cultural practices, traditional view points or to the wide field of human creativity and, as such, to aspects of intangible cultural heritage.

On the other hand, certain objects communicate to us the creative forces of endogenous and exogenous processes or expressions of former life and, by extension, of intangible natural heritage.

I would like to question the opinion common in the human sciences that nothing, including nature, exists free of human interpretation. For one, all human thinking and action is mimetically influenced and, second, we too, through our instincts, are animal-like beings. The resulting question then becomes how are we different from animals or, the other way around: do animals have culture?

For apes this question has long been answered. Chimpanzees *Pan troglodytes*, for example, like humans, live in cultures and differentiate from each other in traditions that, independent from heredity and environment, are being transferred to the following generations. Other animals also provide striking examples: singing birds and their dialects, orcas and their hunting strategies. Even certain ants show the ability to pass down some skills.

Moreover, communication through signs has been observed. Male bowerbirds (*Ptilonorhynchidae*) choose to present themselves in their newly built bowers to the females, making sure that the many decorations, which may include shells, pebbles and twigs in various designs and often in very specific color patterns, do not go unnoticed.

Is this intangible cultural or intangible natural heritage?

As comprehensive intangible natural heritage, one could understand all natural elements that, though untouchable, touch the human senses. Such reality embraces more than what can be transmitted through signs, symbols, text or language. Possibly man, in his ongoing attempts to reach out for this (reality?), produced so many different and varying forms of expression and representations.

I think, there is no better (anti)image than the diorama to underline what an intangible natural heritage is. The diorama is at best a room of illusions. The appearance of the natural is fulfilled as a perfect memory of nature. No breeze moves, no animal stirs and no flower withers. Over hangs the deadly silence of the immobile and, in it, the threat of *an unsouled* nature.

ACKNOWLEDGMENTS

I want to thank Tony Clark and Gerry M. Hoorneman for their contributions to translate my essay into the English language.

NOTES

1. International Council of Museums, accessed February 18, 2011, http://www.icomkorea.org/icom2004/program.htm.
2. ICOM NATHIST Call for Papers, 2004, accessed February 24, 2011, http://www.ggwinter.de/icom/fgc.htm.
3. ICOM NATHIST, general conferences and abstracts; Hans-Albert Treff, "Intangible Heritage: Remarks of a Philistine", (paper presented at the ICOM NATHIST meeting in Seoul, Korea, October 2–8, 2004), accessed February 24, 2011, http://www.ggwinter.de/icom/fgc.htm.
4. Atelier Gillmann, Basel, Switzerland.
5. Work was begun on the idea of the exhibition at the end of the 1990s; *Terra Cognita* was on display from 2001 to 2007.
6. As our oldest exhibition-object giving proof of organic life was a 2.7-billion-year-old stromatolite from Australia, we liked the idea of starting our contemporary art program with a presentation of new Australian art: Garry

Carsley (2004/2005), Tony Clark (2006), Troy-Anthony Baylis and William Yang (2006).

7. Ernst Heinrich Philipp August Haeckel (1834–1919) was an eminent German biologist, philosopher and professor. From 1899 to 1904, he published *Kunstformen der Natur*, "Artforms of Nature", including over 100 plates with detailed, multicolor illustrations of animals and sea creatures.
8. "Exemplary Museum in Essen", Tudo Feist, Beispielgebendes Museum in Essen, *Saarbrücker Zeitung*, 2005 (1–06–02).
9. The exhibition *Down and Under*, designed by Ursula Gillmann, was presented in 2000. Several aspects and new collections of this exhibition are now to be seen on permanent display the new Ruhr Museum, which was opened to public on the January 10, 2010, in one of the biggest and most representative industrial buildings in the Ruhr Area: the coal-washing plant of the Zollverein-Colliery, shaft XII. In an integrative concept, the museum presents the natural history as well as the cultural history of the Ruhr Area.
10. Siegfried Saerberg, "Klangraum Ruhrgebiet", in *Unten und Oben. Die Naturkultur des Ruhrgebiets*, ed. Ulrike Stottrop (Bottrop: Peter Pomp, 2000), 46–47.
11. Now called the Ruhr Museum.
12. Ursula Gillmann, "Notizen zur Ausstellungsgestaltung", in *Unten und Oben. Die Naturkultur des Ruhrgebiets*, ed. Ulrike Stottrop (Bottrop: Peter Pomp, 2000), 19.
13. The exhibition was a joint venture between *Ludwig Galerie Schloß, Oberhausen and Ruhrlandmuseum Essen*, the collection Ann und Jürgen Wilde, Zülpich, and the *Alfred-Ehrhardt-Stiftung, Köln/Berlin* and was curated by Peter Pachnicke and Ulrike Stottrop in collaboration with Udo Scheer and Bernhard Mensch.
14. Albert Renger-Patzsch, *Gestein. Photographien typischer Beispiele von Gesteinen aus europäischen Ländern. Mit einer Einführung von Max Richter und mit einem Essay von Ernst Jünger* (Ingelheim am Rhein: C. H. Boehringer Sohn, 1966).
15. Christine Stahl, "Albert Renger-Patzsch und Alfred Ehrhardt: Metamorphose der Steine", in *Living Stones. Die Natur als Künstlerin,* eds. Bernhard Mensch, Peter Pachnicke and Ulrike Stottrop (Bottrop: Peter Pomp, 2007 [updated 2009]), 192.
16. Ernst Kalai, 1927, cited in Christine Stahl, "Albert Renger-Patzsch und Alfred Ehrhardt: Metamorphose der Steine," 192.
17. Karl Blossfeldt was a German photographer, sculptor, teacher and artist who worked in Berlin, Germany. He is best known for his close-up photographs of plants. We showed a selection of his fascinating photography.
18. Christine Stahl, (full text, see n15) "Albert Renger-Patzsch und Alfred Ehrhardt: Metamorphose der Steine", 194–195.
19. Gen. 9:2, King James version.
20. Hua Wang Lü, 16th century.
21. Robert D. Mowry, *Chinesische Gelehrtensteine aus der Sammlung von Richard Rosenblum* (Zürich: Museum Rietberg, 1998), 10.
22. Albert Lutz and Alexandra von Przychowski (eds.), *Wege ins Paradies oder die Liebe zum Stein in China. Felsen und Steine in der chinesischen Malerei* (Zürich: Museum Rietberg, 1998), 10.
23. M. Catherine Fagg, "Rock music", *Occasional Paper on Technology* 14, (1997).
24. *Ibid.*
25. A Korean stand of chime stones is, certainly, one of the masterpieces of intangible heritage.

26. M. Catherine Fagg, "Rock music", *Occasional Paper on Technology* 14, (1997). 3.
27. Cited in James Blades, *Percussion Instruments and Their History* (London: Faber and Faber, 1975), 90–91.
28. M. Catherine Fagg, "Rock music", *Occasional Paper on Technology* 14, (1997). 3.
29. James Blades, *Percussion Instruments and Their History*, 90–91.
30. Fine, medium and coarse-grained stones can be seen to resonate. The most important factor seems to be the crystalline structure. The position of the ringing rock is also important for increasing or damping the sound. It must be as free as possible to be able to vibrate.
31. *Sound of Stones* is still on display in the Mineral Museum Essen.

9 Discussion

Towards a Unified Concept of Intangible Natural Heritage

Eric Dorfman and Janet Carding

WHAT HAVE WE LEARNED FROM THE CHAPTERS IN THE BOOK?

In the Introduction (Chapter 1), Eric Dorfman posed a question around the nature of intangible natural heritage. The contributors to this book have presented a wide range of approaches to the subject, exploring its depth and breadth. Although, despite their apparent disparity, a surprising number of commonalities emerge. In this final chapter, we use these approaches to develop a unifying concept that both binds and evaluates the utility of these perspectives to develop an overarching, useable synthesis.

The counterpart to our subject, intangible cultural heritage has, over the last few years, gained considerable momentum as a term and a topic for discussion. As it has done so, institutions around the world have begun to survey and take steps to protect it.[1] When that process of mapping began, a definition of intangible cultural heritage was needed to frame and focus the endeavor. For instance, in their 2008 survey of Scottish intangible cultural heritage, the team at Napier University quoted from the intangible cultural heritage context around the definition contained in the United Nations Educational, Scientific and Cultural Organization (UNESCO) convention.

> *The 'intangible cultural heritage' means the practices, representations, expressions, knowledge, skills—as well as the instruments, objects, artifacts and cultural spaces associated therewith—that communities, groups and in some cases individuals recognize as part of their cultural heritage. This intangible cultural heritage, transmitted from generation to generation, is constantly recreated by communities and groups in response to their environment, their interaction with nature and their history, and provides them with a sense of identity and continuity, thus promoting respect for cultural diversity and human creativity. For the purposes of this Convention, consideration will be given solely to such intangible cultural heritage as is compatible with existing international human rights instruments, and complies with the requirements of mutual respect among communities, groups and individuals of sustainable development.*[2]

This description serves to put boundaries around the concept as involving inheritance from previous generations, describing activities happening at a group level rather than to an individual and being compatible with principles of human rights and sustainability.

An analogous definition of intangible natural heritage might simply be to replace the word "natural" for "cultural" in the definition above.[3] This seems, however, to raise immediate problems. For instance, when intangible cultural heritage describes human activities, pointing to previous generations and human rights seems appropriate, but when speaking of intangible natural heritage, the situation becomes less clearly definable. Describing "natural heritage" as the legacy the natural world has inherited from previous generations categorizes it according to a purely human dimension of measurement. For a world that, by recent estimates, is around 4 billion years old and has had modern humans for just 200,000 years or so, judging nature using human benchmarks seems to exclude more that it includes. To remove the human element, one might instead talk of a phenomenon that has existed for a period of time and still exists, in some form, in the present.

Alternatively, taking natural heritage as a starting point might be more fruitful. If natural heritage encompasses the geological, botanical, zoological and meteorological phenomena that make up the environment around us, then surely intangible natural heritage becomes the immaterial attributes of these phenomena. By this reasoning, intangible natural heritage would be, as Ulrike Stottrop (Chapter 8) discusses, the noise of a river eroding its surrounding rocks, the smell of a pine forest, the call of a bird and the whistle of the wind. While named and described by humans, these attributes are not dependent on us for their existence and, while individual circumstances have changed, could all have said to have existed over a period of time. Thus, this grouping, which we suggest is termed "naturogenic intangible natural heritage" could be seen as one of the logical subgroups of the phenomenon. However, examining an example of naturogenic intangible natural heritage suggests that while these phenomena certainly are intangible and natural, it is in their interrelationship with humans that they can be seen as having heritage value. This point, in one way or another, is made by all the authors and, indeed, Stottrop, as well as Jin and Yen, suggest that only a small component of the relationship to natural objects or systems might be about their materiality.

> *To observe nature, to experience, describe and analyze it with scientific methods and to dissect it, is a practice formed more by the cultural continuum of the observer than by the subject of observation itself.*[4]

The term "naturogenic" distinguishes the concept from its corollary, "anthropogenic intangible natural heritage", in which humans create the heritage in the form of traditions focused on nature. We explore the context of these subdivisions in the examples below.

FINDING MEANING IN THE INTANGIBLE—THE SMELL OF HORSES

Horses use scent to communicate with one another through pheromones. Odorless but detected through the olfactory organs, these species-specific chemicals are used in different ways. For instance, nursing mares, like all lactating mammals, release substances called appeasing pheromones (in this case, "equine appeasing hormone"—EAP) to provide reassurance to their offspring, especially in unknown situations, and to aid in bonding between the foal and its mother. Horse trainers are able to connect into this phenomenon for their benefit, by using synthetic EAP as a nasal spray to calm adult horses in potentially frightening situations.[5] However, despite this convenience, horses' olfactory relationships happen completely in the world of the horse, bearing no direct impact on humans culturally or psychologically.

Similarly, horses use pheromones competitively to mark their territories, and repeat visitation to the same small area for elimination causes closed fields grazed by horses to develop a typical pattern of areas of short, grazed grass and others of tall, ungrazed grass, used for waste. This act of competition among individual horses has an important economic impact in areas with intensive grazing, due to loss of nutritive surface of paddocks,[6] and helps explain equine social behavior. Nevertheless, it seems difficult to justify it as heritage. It is only when the scent of horses begins to intersect with the human world that the phenomenon begins to take on heritage value. In our day-to-day lives, our 'olfactory consciousness' helps us enrich our relationships and the sense of our environment. People have many different reactions to the scent of a horse, and for those who have positive associations with horses, the scent triggers a highly favorable response.

> *I learned to love that distinct smell of horse as I buried my face in their soft coats. As clear as if they were talking they were telling me things would be alright, and no matter what . . . they would love me.*[7]

Despite our strong associations with the scent, it cannot be said that horses are communicating with us using olfaction, even passively. We ourselves form associations with the smell based on personal experiences, and the horse contributes only by existing, and being led by its rider into shared experiences. This is the most obvious component of an olfactory natural heritage, triggered whenever we smell a horse and, perhaps, subtly influencing our longer-term responses when we see or even hear about them.

This simple example of the scent of a horse illustrates how intangible natural phenomena only become intangible natural *heritage* when being described in relation to the human experience. As with Xingbao Jin and Alan Yen's discussion of the relationships people in China form to singing crickets (Chapter 7), or even Lee Davidson's exploration of climbers' connection to the New Zealand landscape, it is our association with natural heritage that creates the stories that form the basis of our long-term relationships. By giving these stories a positive value, we make them part of a legacy we wish to preserve and therefore term them 'heritage.' Hence, the concept of intangible natural heritage, like its cultural counterpart, turns out to be inextricably linked to human values, a point alluded to in terms of a variety of natural objects, especially by John Long (Chapter 3) and Markman Ellis (Chapter 4) in discussing, respectively, the emblematic use in Australia of a fossil fish and the kangaroo.

Even while it is possible to conceive of phenomena such as animal behavior or meteorological sounds that have value in their own right, 'value' itself is a human construct. A further example of intangible natural heritage shows how descriptions of natural phenomena quickly become bound to value.

INDIGENOUS KNOWLEDGE AND UNDERSTANDING OF NATURAL HERITAGE

Traditional knowledge of the natural environment is one of the important intersections between intangible natural and cultural heritage and, for many, valued as the fundamental ingredient of the former. Bargaviamma Venugopal (Chapter 5), for instance, considers the 17th-century botanical compendium *Hortus Malabaricus* an "ideal example" of the field, because it taps into traditional knowledge.

Indigenous peoples with a long-term continuity of extracting resources from their local environment often have a deep understanding of the complex behavior of ecosystems, garnered from the slow accumulation of observations, passed down as oral history from generation to generation. Their understanding of the relationship between biodiversity and the provision of environmental goods and services,[8] as well as a longitudinal, heuristic approach to acquiring knowledge about the environment, has led to suggestions that the approach has much to offer the synchronic methodology of western science and, indeed, is vital to effective management of natural resources.[9]

However, in the modern world, values are changing. Many bush skills are no longer essential for livelihoods, and traditions are being incompletely passed down from older generations, or they are lost altogether. Cultural traditions have been the subject of progressive thinking about protection and intellectual property.[10] While progress is still slow, governments and community groups are becoming aware of the loss of capacity in the natural sphere and are beginning to institute measures to counteract these trends. For example, in two subarctic Omushkego (West Main) Cree Indian communities, Moose Factory and Peawanuck, Ontario, Canada, the institution of a hunters' income security program has provided incentives for family units to go onto the land, rather than all-male hunting parties.

This facilitates transmission of women's indigenous knowledge and bush skills, including setting snares and fishnets, beadwork, smoking geese and tanning moose and caribou hides.[11]

In some cases, traditional knowledge contributes directly to wider conservation and economic efforts. In the Northern Territory, Australia, ongoing conservation of saltwater crocodiles is dependent on the economic viability of farming them for meat, skins and blood for research. Northern Territory Crocodiles are usually kept on 'ranches,' in which eggs are harvested from the wild to be hatched on farms and raised to harvesting size (about 1.5 m).

Collecting eggs is difficult. Adults can reach lengths greater than 6 m, are fiercely territorial, with a formidable array of predatory adaptations, and nesting females are particularly aggressive. The hunter-gatherer lifestyle of traditionally living Aboriginal people in the Northern Territory has awarded them with considerable wilderness skills and knowledge around wetlands and crocodiles. The farming industry's demand for eggs collected from the wild provides a rare opportunity to use these skills in this economically depressed region. Despite calls for further development of the industry to involve Aboriginal people more effectively,[12] egg collecting gives local communities about $20 apiece, providing up to 80% of eggs to some farms,[13] and the presence of the industry has facilitated conservation measures that have seen the crocodile population stabilize and grow.

This is clearly intangible heritage. The question is whether it is natural or cultural, given that many people living close to the land reference the wilderness in much that they do. In the example above, we suggest that the relationship of the egg collectors to the crocodiles: their knowledge of animal behavior, the seasonal rise and fall of flood waters and a host of other detail, is firmly intangible natural heritage. The Aboriginal song traditions used to transmit this knowledge are the nexus between intangible natural and cultural heritage.

> *Knowledge of the changes in seasons is land-based and is owned. Finely detailed, such knowledge rests on the relationships between the living things within a particular area. People of the middle Victoria River, for example, know that when the march flies are biting, fresh water crocodiles are laying their eggs. They do not need to go to the places where crocodiles are laying their eggs in order to check up on their progress; march flies 'tell' them when this happening. Further south, however, people know that the crocodiles are laying their eggs because the* Jangarla *tree (*Sesbania formosa*) is flowering . . . This kind of knowledge, local, accurate, and pertinent to people's concerns and interests, is found everywhere in Aboriginal Australia. It is transmitted from one generation to the next in the course of people's daily lives. Such knowledge is also coded in and taught through songs. Some are beautiful little pieces which convey one piece of information but which evoke a whole season and a set of places, people and activities.*[14]

POPULAR PHENOMENA—TULIPMANIA AND THE NATURE OF HERITAGE

If intangible natural heritage is the combination of natural phenomena and human value, then an example of humans explicitly placing value on items in the natural world might further serve to illustrate what we provisionally described as anthropogenic intangible natural heritage and start to indicate the boundaries within which something can be seen as having heritage value.

Perception is a fundamental component of intangible natural heritage. Emotional attachment to an object or subject drives much of our

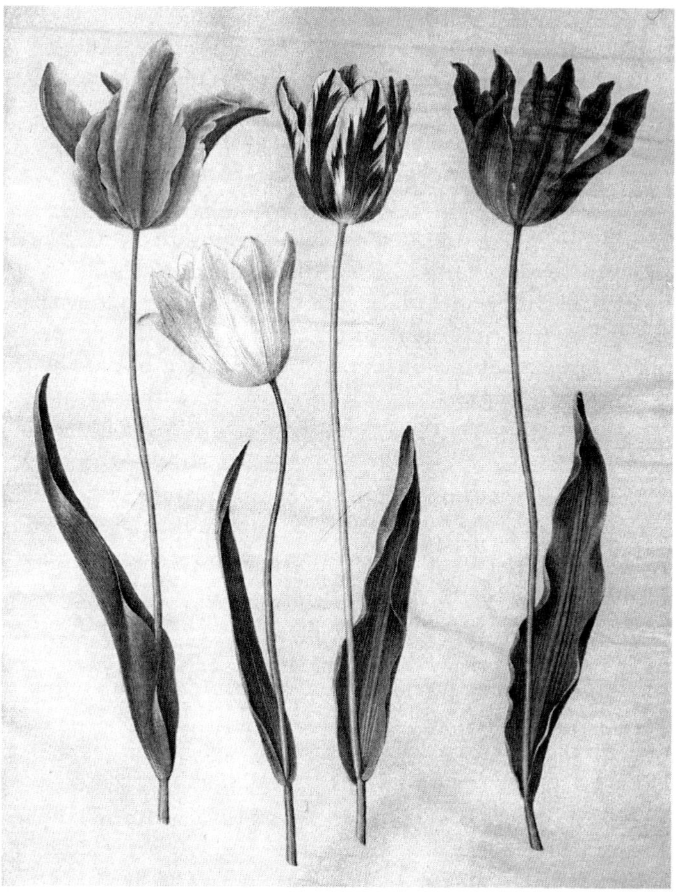

Figure 9.1 *Tulipa gesneriana*, gouache on vellum, from the Gottorfer Codex 1649–1659, a four-volume plant atlas by Hans-Simon Holtzbecker. The illustrations show the effect of TBV in creating various forms of shape and pattern.

ongoing association with it, even long after the initial thrill of novelty has dissipated. In fact, it becomes 'heritage' only when this association is transmitted over time. "Tulipmania" is a famous example, albeit probably misrepresented in popular understanding. The flower *Tulipa* spp., introduced to Europe from the Ottoman Empire in the mid-16th century, quickly became popular, especially in the Netherlands. After 1593, when Flemish botanist Charles de l'Écluse established the tulip in the *Hortus Academicus* at the University of Leiden,[15] it grew quickly into a coveted luxury item and status symbol.

Tulips' reproductive biology, which necessitated labor-intensive propagation, pre-adapted them to becoming objects of desire. The profusion of varieties that were ultimately developed were made more interesting and valuable by the introduction of a tulip-specific mosaic virus known as the "tulip breaking virus" (TBV). Presence of TBV gave the petals of infected plants the sought-after variegated appearance, and can only be passed on through a laborious process of cloning bulbs, made slower still by the virus itself (Figure 9.1). Thus, the most famous varieties were the product of years' exacting work and planning. Growers, naturally proud of their results and eager to reap the rewards, ennobled their new varieties with lofty titles like *Admiral van der Eijck*, *Viceroy* and *Semper Augustus*.

Tulips only bloom for about a week during April and May and from June to September dormant bulbs can be uprooted and moved. It was during these times that speculation took place in what was, essentially, a futures market. Although this practice had its critics at the time, tulipmania only entered popular culture in 1841, with a book[16] by the Scottish journalist Charles Mackay. According to Mackay, the wealth potential of tulips in the early 17th century aroused the interest of the entire nation "even to its lowest dregs", citing outrageous sums of money paid, and goods exchanged, for individual bulbs. People purchased bulbs at increasingly higher prices, with the intention of reselling as the market price rose, hitting a frenzied peak in autumn of 1636. However, this scheme was only effective if, at the end, a buyer could be found to purchase the bulbs and keep them. In Mackay's account, the population of willing end-buyers became exhausted in February 1637, and the whole market abruptly collapsed, sending many to ruin and suicide, and the entire nation into economic chaos.

Mackay used this story as an example of inexplicable society-wide mania, which has, over the intervening time, inspired popular imagination and, largely unchallenged, has been novelized, set to music and analyzed by economists thinking about bursting economic bubbles. Despite Goldgar[17] demonstrating that the phenomenon was in fact restricted to a relatively small group of propagators and aficionados, our fascination with the topic has influenced our relationship with the flower as much as it has our ideas about society-wide fads. It also gives us an important window into this period of European history.

> *This was a new world, a new country, with a new kind of culture, a new set of priorities, a new social system, a new impetus to its commerce and its society. Tulipmania opens a window for us into that culture and its values . . . It might not have been a financial crisis, but it was a social and cultural one. In tulipmania, Dutch burghers confronted a series of issues that in any case gripped their culture: novelty, the exotic, capitalism, immigration, the growth of urban societies, and all the problems and excitement such issues raised. People in the 1630s and after found tulipmania a wonder, something to be marveled at, like a fireball, a child with two heads, or a plague of mice.*[18]

Thus, it is possible that a relatively minor economic event took on a life of its own as a morality tale. Hendrik Van Rheede, author of the seminal *Hortus Malabaricus*, (see Venugopal's description, Chapter 5) was born in 1636, the year that interest in tulips was at its height, and, one must assume, these events played a contextual role in his future awareness of the economic potential of exotic plants.

Every age has had biological phenomena that have captured the hearts and minds of the public. Ostrich eggs and nautilus shells were considered valuable enough to be the bases of gold and silver trophies in the 15th and 16th centuries, and the first goldfish in France (so the story goes) was given by Louis XV to his mistress Madame Pompadour. This passion for novel natural history resonates with the response of Londoners to the arrival of kangaroos from Australia, as described by Markman Ellis (Chapter 4). Today, we have discovered most of what there is to be seen on Earth, except for life in the deep oceans, which still inspires public excitement and the thrill of the unknown.[19] The extreme popularity of a preserved colossal squid at the Museum of New Zealand/Te Papa Tongarewa is case in point.[20]

These various relationships with novel species differ fundamentally from indigenous knowledge, because they are not passed on intergenerationally. In the 21st century, the species we value because they are, like the Giant Panda (Dorfman), highly endangered, like the Passenger Pigeon (per Long's discussion), extinct or because, like the Snowy Owl and Clown Fish, they have been anthropomorphized by Hollywood blockbusters. Like the steady progression of new technology that has, to some extent, taken the place of novel wildlife, their popularity is of the moment, fueled by the lack of availability. Although the massive popularity of tulips in their day may have left significant tangible and intangible cultural heritage (such as the still life works of Johannes Bosschaert and others), it seems difficult to argue a heritable relationship with the tulip as an object of nature on the same level as traditional hunting or even the mountaineering discussed by Davidson. Topics such as this might be as usefully described within the purview of human geography or history as intangible natural heritage. It is only when that thrill of the moment creates a long-lasting, society-wide relationship (such as the kangaroo as synecdoche for Australia, as described by Ellis or, perhaps even more closely, the kiwi *Apteryx* spp. for New Zealand) that it seems appropriate to term it intangible natural heritage.

CONTROVERSIAL TRADITIONS: FOX HUNTING AND BULLFIGHTING

Even those who have no particular association with the wilderness can have strong, even ritualized associations with nature. Following on from indigenous knowledge are the traditions that form around more oblique associations with the natural world. Some activities are seen as wholly positive, like the aesthetic and artistic traditions that form around cricket singing (Jin and Yen), whereas others are generally perceived as negative by all but their practitioners and adherents.

Two prominent examples are fox hunting and bullfighting. Both activities have many similarities: well-defined traditions that stretch back for millennia and became entrenched as modern cultural activities during the late Renaissance (Figure 9.2). Both have highly ritualized customs requiring

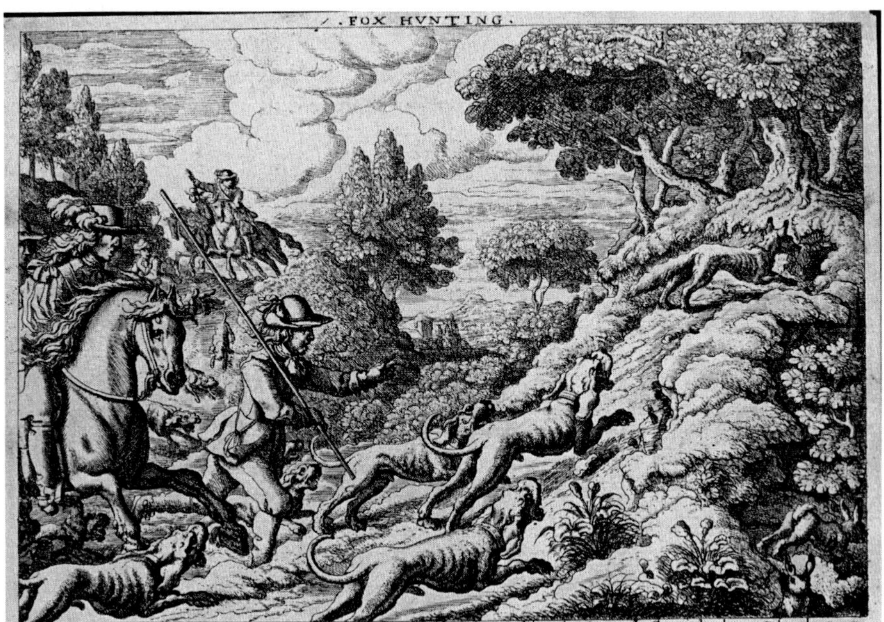

Figure 9.2 "Fox Hunting" by the Czech draughtsman and engraver, Wenzel Hollar (1607–1677). Collection of the University of Toronto, plate number: P2040. Despite a long, tradition-filled history of fox hunting, ethical issues exclude it from our definition of intangible natural heritage.

a close cultural association with the animal, considerable knowledge of animal behavior and skill on horseback. Both are also marked by long histories of controversy[21].

Proponents of fox hunting see it as an important part of rural culture, as well as useful for reasons of conservation and pest control, while the advocates of bullfighting view it as a quintessentially Spanish art form, on a par with flamenco dancing and paella.

Opponents of both argue that it is unnecessarily cruel, pointing out the extreme stress and suffering the animals undergo before they die. Both social rituals and the controversy around animal rights can be seen as 'intangible' and 'natural heritage.' In both activities, body parts of the prey are traditionally awarded to the victor as trophies. A traditional practice in fox hunts was to cut off the "brush" (tail), "pads" (feet) and "mask" (head) as awards, with the carcass then thrown to the dogs. In bullfighting, if the matador has performed particularly well, the crowd may wave white handkerchiefs to petition the event's presider to award him one or both ears, or even the tail.

Modern sensibilities and the animal rights movement have condemned both these sports, which are defended by vocal and influential minorities who argue that they are quintessential parts of national culture. At the time of writing, fox hunting has been banned in Scotland since 2002 and for England and Wales since 2004. In Spain, bullfighting is no longer televised nationally.

At the start of this chapter, the UNESCO definition of intangible cultural heritage included criteria such that only those practices that were consistent with human rights and sustainability could be seen as worthy of preservation as intangible cultural heritage. We suggest that fox hunting and bull fighting reveal analogous criteria of legality, community support, ethical treatment of animals and environmental sustainability, for intangible natural heritage. Those making a claim of intangible natural heritage worthy of preservation would need to mount a convincing case that the natural phenomena and the value of their interrelationship with humans was consistent with local laws and would not result in the ending or extinction of the animal, plant or natural object. The examples here represent instances where human impact means that we believe these conditions are not met.

CONSUMPTIVE AND NON-CONSUMPTIVE USE

Carrying the theme of controversy further, intangible cultural and natural heritage can clash when the results of cultural activity are detrimental to wildlife and landscape. Ironically, the largest impact on species often stem from the very features for which they are admired. This is another aspect of the negative (or dangerous) side of intangible natural heritage.

Rhinoceros Horn in Traditional Chinese Medicine

Adrian Norris (Chapter 2) discusses the ethics of museums releasing collection items that have the potential to fuel black market trading. Rhinoceros horn has been used in traditional Chinese medicine since at least 200 BC. However today, with the population of adherents of these practices skyrocketing, the impact of this demand (coupled with high demand from Yemen for horn used to make cultural artifacts) is sending the species extinct. In the 17 years between 1970 and 1987, the world's rhinoceros population plummeted from 70,000 to 11,000, a drop of 85%,[22] and, where supply had dried up due to national and international legislation, it continues through the black market. This precipitous decline will continue unless effective conservation programs can be launched to protect these animals from illegal poaching.

The Convention on International Trade in Endangered Species of Wild Fauna and Flora (CITES) has listed all five species of rhinoceros in Appendix I, prohibiting these animals or their products from commercial trading in and among states that are signatories of CITES. Additionally, the governments of many Eastern countries and special districts, including mainland China, Taiwan, Hong Kong, Macao, Singapore, Japan, South Korea and Thailand, have enacted legislation to ban export and import of rhinoceros horn. However, this does little to alter the deeply ingrained beliefs of the practitioners of Chinese traditional medicine, where the material is considered important as a cure for conditions associated with "excessive heat in the blood" (antipyretic), and thus relieving a wide variety of ills including toxicosis, febrile diseases, influenza, high fever, poisoning, convulsion, epilepsy, restlessness, delirium, macular eruptions, erysipelas, hemoptysis, epistaxis, carbuncle, malignant swelling and abscess.[23] It has even been postulated to possess a cure for AIDS,[24] although it is not, contrary to popular perception, used as an aphrodisiac or in promoting virility.[25]

In an effort to aid conservation efforts, But *et al.*[26] in 1990 compared the effectiveness of rhinoceros horn in reducing fever with horn from water buffalos *Bubalus bubalis* and, perhaps ironically, the saiga antelope *Saiga tatarica*, a Eurasian steppe species, itself endangered by poaching for traditional medicine. Under clinical conditions, all three performed equally well, providing the dosage was adjusted appropriately. This research, demonstrating more than two decades ago the potential substitution of the near-ubiquitous water buffalo, has had no impact on the demand for medicinal rhinoceros horn and poaching, facilitated by sophisticated weaponry, continues to increase steadily.[27] Why should the demand continue when perfectly acceptable, environmentally sustainable and undoubtedly cheaper options exist? Government lapses, private sector greed and Asian crime syndicates may explain what supplies the market, but not the psychology that sustains it. This profoundly destructive connection to the species must be understood before truly effective regulation can be imposed. However, this relationship does not meet the criteria for intangible natural heritage.

Whale Watching and Ecotourism

Norris mentions British leaflets encouraging trampers to "take only photographs and leave only footprints". However, impact is not always easy to measure: tropical corals and alpine lichens are vulnerable to being stepped on and glowworms are sensitive to the camera's flash. However, non-consumptive use of the environment, if such a thing exists, is at least a useful goal.

Aspects of whale watching could, arguably, fit under the umbrella of intangible natural heritage on the grounds that the conservation ethos that people gain in this family-oriented activity are intergenerational, and the industry has added to a long-term change in sensibility around whales and the marine environment. The values people develop from whale watching arguably aligns with Davidson's analysis of wilderness values, "seeing nature on its own terms" and fits firmly within the realm of ecotourism.

The desire of urbanites to reconnect to nature and its spectacle has roots in mountaineering movement of the 19th century, or even the Grand Tour of the 17th century at a time when, as Ellis points out, people were becoming increasingly cut off from the natural world. However ecotourism, formalised into a *bona fide* tourist activity in the 1980s, is fundamentally different. The tenets of ecotourism are that the activity should be nature-based, environmentally educated and sustainably managed. It is often argued that one of the benefits of ecotourism is the potential to protect wildlife through allowing local people to become active participants in the industry. In turn, the distribution of industry benefits should play an important role in facilitating access to wildlife resources for tourism purposes and in ensuring their conservation.[28] In this, ecotourism and intangible natural heritage are well aligned.

As an eco-tourist activity, whale watching has been put under considerable scrutiny, seen variously as a much-needed billion-dollar input to local economies worldwide[29] or a harmful form of exploitation, more subtly impacting than hunting, but still potentially damaging, disrupting social and feeding behaviors, potentially shifting habitat use to less productive areas.[30] The truth of the story is complex and has elements of both. First of all, the relationship between whales and boats is complex. For instance, killer whales *Orcinus orca* in British Columbia appear to use a variety of behaviors to avoid vessels, which depend in part on vessel numbers, proximity and possibly the size, activity and noise of a given vessel. However, individual responses appear to differ. Such is the variability of the system that no specific short-term effects have been quantifiable. The fact that killer whales continue to use their core areas despite high levels of whale watching traffic, suggests that there are no longer term effects either.[31]

However, off Kaikoura, on New Zealand's South Island, where sperm whales *Physeter macrocephalus* are the focus of a boat and aircraft-based whale watching industry, several aspects of whale behavior were significantly affected by the presence of whale watching vessels. Research[32] has demonstrated that whale-watching boats and aircraft, individually or together, cause increases in the time whales spend at the surface and in the frequency and number of changes in direction, and reduce the incidence of aerial behaviors. However, like killer whales, behavior changes with the individual. In this case, where whales are resident, returning year after year, they react less strongly than those who are transient. Additionally, some whales are by nature more tolerant than others. The result of this occasional misalignment between intangible natural and cultural heritage is an ethical issue which, in cases such as so-called "scientific whaling", makes its way to the highest levels of national and international government.

Tying It Together

The examples in the preceding discussion have begun to make visible the boundaries around the concept, and suggest that a notion of intangible natural heritage requires a linking of the intangible phenomena of the natural world with a human interrelationship and includes the natural phenomena being seen as valuable and so worthy of preservation. Although there is clearly much work to be done before it would be possible to put forward a definition as developed as that articulated by UNESCO for intangible cultural heritage, we offer the following draft for consideration:

> *Intangible Natural Heritage –The environmental forces that create biological and geological entities, the phenomena that these entities produce, and their interaction with humans and human communities. Particularly included are those phenomena and interactions that are repeatable, rather than isolated incidents, are compatible with legal frameworks, and are consistent with sustainable development that does not diminish the natural world for future generations.*

So while referring to natural phenomena, intangible natural heritage includes a human dimension as an integral element, without which it is difficult to evaluate whether or not a phenomenon is worthy of preservation. It is this notion of value that enables us to start developing criteria for what might usefully be agreed to be included in, or excluded from, intangible natural heritage. This is because, as many of the authors in this volume point out, the value of nature and, by extension, that of intangible natural heritage, it is in itself a cultural phenomenon.

Although this discussion teeters on the edge of philosophical abstraction, it is worth stating that attributing value to the various forms of heritage in this way is at the heart of the conservation movements that have led to the development of the UNESCO conventions. While society is comfortable with placing human value on Stonehenge or the Taj Mahal, conversations around the natural world have often been around their intrinsic value rather than described in terms of their relationship to humans. However, authors with interests as diverse as Norton from a public policy perspective[33] through to Richard Louv, an advocate of outdoor play for children,[34] have recently urged that nature should be given a positive value based on their worth to us. Interestingly, both Norton and Louv suggest that, since humans will only conserve that which they value, encouraging the valuing of nature is an essential precursor to any protection of the natural world. This is an opinion shared by Jin and Yen, Stottrop and Davidson. From that point of view, activities that connect people with natural phenomena could be seen as a way of creating intangible natural heritage.

FUTURE DIRECTIONS

A volume such as this is necessarily introductory, so it is worth speculating for a moment about fruitful avenues of investigation from this point. A first direction would be to investigate more fully the philosophical discussion that has been initiated by the authors in this volume. Researchers might want to interrogate more fully the notions of human value and heritage touched on here, and in doing so, perhaps either find our definition of intangible natural heritage to be robust or, perhaps more likely, to suggest improvements and additions. A further development down this path might be to explore the ethical linkages between intangible natural heritage and territory such as animal rights and welfare.

Research

A second avenue would be to continue to seek out new examples and case studies of what might be proposed as intangible natural heritage and to investigate its relationship with both tangible natural phenomena but also different cultural settings and communities. As examples put forward in this volume begin to reveal, some of what is put forward here as intangible natural heritage does not easily fall within any preexisting discipline but occupies an interdisciplinary territory between history, geography, ecology and the sociology of natural history. There are, we suggest, useful discussions to be had around the intangible interrelations between the human and natural world at various points in time and location.

Action

Recognition of the importance of everybody's relationships to the Earth and doing things to protect and enhance those relationships is key to the discussions in this book. Just as UNESCO has given stature to intangible cultural heritage, so too could exploration of intangible natural heritage lead to steps to survey and preserve it. Projects that increase awareness of urban biodiversity that include encouraging people to find value in their relationship with the natural world, could also be seen as examples of action to make clear, and gain support for, conserving and developing intangible natural heritage. Examples include Australia's *Birds in Backyards* program, described in the Introduction, the *Woods on Your Doorstep*[35] program of Britain's Woodland Trust or, in the U.S., the Audubon Society's *Audubon Adventures* program.[36]

Curation and Interpretation of the Intangible

The concept of intangible natural heritage, like any theoretical framework, is only as useful as its applicability to people who are going to be using it. Moreover, as Norris points out, considerable challenges face museums into the future, both operationally and philosophically. For natural history museums, which have traditionally made collections from the tangible parts of the natural world, the notion of intangible natural heritage is every bit as testing as the idea of intangible cultural heritage is to museums of anthropology and social history. There have been some very good initial forays into this, although they aren't named as such. Jin and Yen, as well as Stottrop, have described their display of the intangible. Aligned to that, a major traveling exhibition, *Music of Nature,* was initiated in the U.S. in 2007. It covered the mutual interplay between music and nature, as a result of a collaboration an interdisciplinary group of scientists and musicians from the Association of Science-Technology Centers, the Science Museum of Minnesota and the BioMusic Program of National Musical Arts. BioMusic researchers combined the analytical tools of musicology with research in a wide variety of fields: neuroscience, biology, zoology, environmental science, physics, psychology, mathematics and anthropology.

> *Sound technology and cross-disciplinary approaches to research are creating insights into the nature of music that are as yet little known by the general public. This emerging research field was acknowledged in 1986 by the National Academy of Sciences, when the BioMusic Program was established by NAS's resident ensemble, National Musical Arts (NMA). In recognizing the program, NAS's then-president Frank Press referred to "the powerful new understanding being offered by the wider realization of the shared principles of science and art . . . and a renaissance spirit of blending the different parts of knowledge to help us understand the whole".*[37]

Our hope is that interest in the concept of intangible natural heritage will lead to more examples within natural history museums, and indeed elsewhere, that interpret and use intangible natural heritage as a starting point for public programs and collection development.

Language and Interpretation

To 'bed in' such a new concept to the field, and create a new focus is important in that it gives a fresh perspective that can invigorate discussions around conservation of the natural world and human culture. Many aspects of the philosophy have yet to be properly developed, or are 'claimed' by other branches of conservation and social science, museology or public awareness. If intangible natural heritage is to succeed as a concept it, like the term "biodiversity" or the Gaia Hypothesis, must take on a life of its own and enter common usage. In time, the term "intangible natural heritage" might even give way to another phrase more recognizable and relevant to the people using it. Yet even if this occurs, the examples in this volume and these thoughts that build a coherent idea of the concept will, we hope, have been a useful introduction to the notion that there are invisible connections between humans and natural heritage and that their strength and diversity is vitally important for both.

NOTES

1. See, for example, the report commissioned by Museums and Galleries Scotland, Alison McCleery, Alistair McCleery, Linda Gunn and David Hill, *Scoping and Mapping Intangible Cultural Heritage in Scotland Final Report* (Edinburg: Napier University, 2008), accessed on January 16, 2011, http://www.museumsgalleriesscotland.org.uk/publications/publication/71/scoping-and-mapping-intangible-cultural-heritage-in-scotland-final-report.
2. *Ibid.*
3. This was the approach suggested by Hans-Albert Treff, as described by Ulrike Stottrop, Chapter 8.
4. Stottrop, Chapter 8.
5. Christelle Falewee, Emmanuel Gaultier, Céline Lafont, Laurent Bougrat and Patrick Pageat, "Effect of a synthetic equine maternal pheromone during a controlled fear-eliciting situation", *Applied Animal Behaviour Science* 101 (2006): 144–153.
6. Frank Ödberg and Keiper Francis-Smith, "Studies on the formation of ungrazed eliminative areas in fields used by horses", *Applied Animal Ethology* 3 (1977): 27–34.
7. DeLila Lumbardy, "A horse is a horse . . . of course", *The Fence Post,* February 14, 2010, accessed December 2, 2010, *http://www.thefencepost.com/article/20100214/FEATURES/100219990.*
8. Madhav Gadgil, Firket Berkes and Carl Folke, "Indigenous Knowledge for Biodiversity Conservation", *Ambio* 22 (1993): 151–156.
9. Firket Berkes, Johan Colding and Carl Folke, "Rediscovery of traditional ecological knowledge as adaptive management", *Ecological Applications* 10 (2000): 1251–1262.
10. See, for instance, Dennis S. Karjala and Robert Kirkwood Paterson, "Looking beyond intellectual property in resolving protection of intangible cultural

heritage of indigenous peoples", *Cardozo Journal of International and Comparative Law* 11 (2003): 633.
11. Kayo Ohmagari and Firket Berkes, "Transmission of indigenous knowledge and Bush skills among the Western James Bay Cree women of subarctic Canada", *Human Ecology* 25 (1997): 197–222.
12. Graham Webb, Charles Missi and Miriam Cleary, "Sustainable Use of Crocodiles by Aboriginal People in the Northern Territory", in *Sustainable Use of Wildlife by Aboriginal Peoples and Torres Strait Islanders,* eds. Mary Bomford and Judy Caughly (Canberra: Australian Government Publishing Service, 1996): 167–173.
13. Clem Tisdell, Clevo Wilson and Hemanth Swarna Nantha, *A Report on the Management of Saltwater Crocodiles (Crocodylus porosus) in the Northern Territory: Results of a Survey of Pastoralists.* Working Paper on Economics, Ecology and the Environment 130 (Brisbane: University of Queensland, 2005), 11.
14. Deborah Bird Rose, *Nourishing Terrains: Australian Aboriginal Views of Landscape and Wilderness* (Canberra: Australian Heritage Commission, Canberra, 1996).
15. Mike Dash, *Tulipomania: The Story of the World's Most Coveted Flower, and the Extraordinary Passions It Aroused* (London: Gollancz, 1999).
16. Charles Mackay, *Extraordinary Popular Delusions and the Madness of Crowds*, with foreword by Andrew Tobias (New York: Harmony Books, 1980; originally published 1841).
17. Anne Goldgar, *Tulipmania: Money, Honor, and Knowledge in the Dutch Golden Age* (Chicago: The University of Chicago Press, 2007).
18. *Ibid.*
19. Data from the Australian Museum.
20. Data from the Museum of New Zealand/Te Papa Tongarewa.
21. For instance, in 1567, Pope Pius V issued a papal bull banning bullfighting on the grounds of cruelty, which was repealed by Gregory XIII in 1575 at the request of Phillip II of Spain. Burns and Oates, *The Church and Kindness to Animals* (Whitefish: Kessinger Publications, 2007; originally published in London, 1906).
22. Esmond B. Martin and Lucy Vigne, "Recent developments in the rhino horn trade", *Traffic Bulletin* 9 (1987): 49–53.
23. Hong-yen Hsu, Yuh-pan Chen, Shuenn-jyi Shen, Chau-shin Hsu, Chien-chih Chen and Hsien-chang Chang, *Oriental Materia Medica, A Concise Guide* (Long Beach: Oriental Healing Arts Institute, 1986).
24. B. P. Wu and S. K. Lu, "Discussions on the possible treatment of AIDS with traditional Chinese medicine", *Beijing Journal of Traditional Chinese Medicine* 1 (1986): 14–16.
25. Esmond Bradley Martin, *Rhino Exploitation, the Trade in Rhino Products in India, Indonesia Malaysia, Burma, Japan and South Korea* (Hong Kong: World Wildlife Fund Hong Kong, 1983).
26. Paul Pui-Hay But, Lai-Ching Lung and Tam Yan-Kit, "Ethnopharmacology of rhinoceros horn. I: antipyretic effects of rhinoceros horn and other animal horns", *Journal of Ethnopharmicology* 30 (1990): 157–168.
27. Tom Miliken, Richard H. Esmile and Bibhab Talukdar, *African and Asian Rhinoceroses—Status, Conservation and Trade* (a report from the International Union for Conservation of Nature and Natural Resources Species Survival Commission (IUCN SSC) African and Asian Rhino Specialist Groups and TRAFFIC to the CITES Secretariat pursuant to Resolution Conf. 9.14 (Rev. CoP14) and Decision 14.89, 2009).

28. Clem Tisdell, *Sharing Nature's Wealth through Wildlife Tourism: Its Economic, Sustainability and Conservation Benefits.* Working Papers on Economics, Ecology and Environment 170 (Brisbane: School of Economics, University of Queensland, 2010).
29. Simon O'Connor, Roderick Campbell, Hernan Cortez and Tristan Knowles, *Whale Watching Worldwide: Tourism Numbers, Expenditures and Expanding Economic Benefits,* a special report from the International Fund for Animal Welfare (Yarmouth, MA: Economists at Large, 2009).
30. Mark B. Orams, *Marine Tourism: Development, Impacts and Management* (London: Routledge Press, 1999).
31. Rob Williams, Andrew Trites and David E. Bain, "Behavioural responses of killer whales (*Ornicus orca*) to whale-watching: opportunistic observations and experimental approaches", *Journal of Zoology* (London) 256 (2002): 255–270.
32. Christoph F. Richter, Stephen M. Dawson and Elisabeth Slooten, "Sperm whale watching off Kaikoura, New Zealand: effects of current activities on surfacing and vocalisation patterns", *Science for Conservation* 219 (2003): 5–78.
33. Bryan G. Norton, "Sustainability Indicators: Integrating Evaluation Studies into Adaptive Management Indicators: Integrating Evaluation Studies into Adaptive Management", in *Forum on Sustainability, Well-Being, and Environmental Protection: What's an Agency to Do?* (EPA Economics Forum, Washington, D.C., December 2, 2005), accessed February 22, 2011, http://www.epa.gov/sustainability/pdfs/norton-epa-ord-paper.pdf.
34. Richard Louv, *Last Child in the Woods: Saving our Children from Nature-Deficit Disorder* (Chapel Hill: Algonquin Books, 2005).
35. Accessed February 19, 2011, http://www.woodlandtrust.org.uk/en/about-us/projects/woyd/Pages/woyd-uk.aspx.
36. Accessed February 19, 2011, http://education.audubon.org/audubon-adventure.
37. Association of Science and Technology Centers (U.S.), accessed February 19, 2011, http://www.astc.org/resource/exhibits/sound/biomusic.htm.

Contributors

Janet Carding: Director and Chief Executive Officer (CEO), Royal Ontario Museum (ROM), Canada. Janet Carding has been the Director and CEO of the ROM since September 2010. She began her career as a curator at the Science Museum in London, England, prior to focusing on exhibition and programs development. Carding's scope of work includes managing projects during construction of the Science Museum's Wellcome Wing extension, leading exhibition development teams, and coordination of a review of the National Museum of Science and Industry (NMSI), London, resulting in new strategic direction and organizational changes. Prior to joining the Australian Museum, Carding was Head of Planning and Development with NMSI, leading strategic planning and major projects for NMSI's three museums. She joined the ROM from the Australian Museum, Sydney, the country's preeminent museum of nature and culture, where she has held the position of Assistant Director, Public Programs and Operations since 2004.

Lee Davidson: Senior Lecturer, Department of Museum and Heritage Studies, Victoria University of Wellington, New Zealand. Dr Lee Davidson has been teaching the Museum and Heritage Studies program and its predecessor, Recreation and Leisure Studies, since 1998. Previously she worked in various positions within the recreation, tourism and conservation sectors. Her research interests include leisure (history, theory and contemporary practice), visitor studies, narrative research methods and tourism and natural/cultural heritage. She is a board member for the Australia and New Zealand Association for Leisure Studies and was the recipient of their Thesis of the Year Award in 2007. She is currently writing a book with Robert A. Stebbins entitled *Serious Leisure and Nature* (contracted with Palgrave/Macmillan), and she is working with Museums Aotearoa to develop a national framework for museum visitor research in New Zealand.

Eric Dorfman: Director, Whanganui Regional Museum, New Zealand. Eric Dorfman is Director of Whanganui Regional Museum and has a PhD in wildlife ecology (University of Sydney), a Masters in marine biology from San José State University (SJSU) and is an Adjunct Lecturer at Victoria University of Wellington in the Department of Museum Studies. He is also Vice President of the International Council of Museums' Committee for Museums and Collections of Natural History (ICOM NATHIST). Credits include the CD *Wild Music*, co-produced with Radio New Zealand Concert and performed by the New Zealand Symphony Orchestra, as well as the documentary *Te Kereru: The World of the New Zealand Pigeon* (writing and direction) for the Museum of New Zealand, narrated by Dame Kiri Te Kanawa. His published works include short fiction and the popular science book, *Melting Point: New Zealand and the Climate Change Crisis*, published by Penguin New Zealand.

Markman Ellis: Professor of 18th-Century Studies School of English and Drama, Queen Mary, University of London, United Kingdom (UK). Markman Ellis's research concerns 18th-century literature and culture in English. His recent research has focussed on the representation of the coffee-house in the period 1650–1750, with a monograph entitled *The Coffee-House: A Cultural History* published by Weidenfeld and Nicolson in November 2004. He has recently completed a four-volume facsimile edition of coffee-house satires (1657–1780) for Pickering and Chatto. His first book was a study of political controversy in sentimental novels, entitled *The Politics of Sensibility* (Cambridge University Press, 1996, now in paperback), which developed an argument about the feminization of culture in 18th-century Britain. He has also published T*he History of Gothic Fiction* (Edinburgh University Press, 2000) and co-edited Discourses of Slavery and Abolition (Palgrave, 2004). Other topics within 18th-century studies that he has addressed in articles and chapters include: georgic poetry and ideas of empire; travel writing and the rhetoric of wonder; slavery and sensibility. His current research is a project on what it means to be a critic in the early 18th century, entitled "The Social Space of Criticism".

Xingbao Jin: Chief Editor of "Nature and SciTech" Publishing House, Shanghai Science and Technology Museum, China. Xingbao Jin graduated from Fudan University in 1975 and received an MS in Systematics of Entomology from the Chinese Academy of Sciences in 1981 and a PhD in Systematics of Entomology from McGill University, Montreal, Canada, in 1991. Jin was a senior researcher systematist in orthoptera insects, with a specialty in katydids and crickets, the singing insects. Jin served as Deputy Director of the Shanghai Natural History Museum from 1997 to 2000 and Deputy Director of Shanghai Science and Technology Museum since 2000 when the two museums were merged. Now Jin is in charge of education and foreign affairs in the museum. Research interests include urban nature conservation and museology since 1997.

Jin has published several papers and a few books on the above-mentioned fields.

John A. Long: Vice President, Research and Collections, Natural History Museum of Los Angeles County, U.S.. John A. Long is an Australian paleontologist who is currently the Vice President of Research and Collections at the Natural History Museum of Los Angeles County. He is also an author of popular science books. His main area of research is on the fossil fish of the Late Devonian Gogo Formation from northern Western Australia. It has yielded many important insights into fish evolution, such as *Gogonasus* and *Materpiscis*, the later specimen being crucial to our understanding of the origins of vertebrate reproduction. His love of fossil collecting began at age seven, and he graduated with a PhD from Monash University in 1984, specializing in Palaeozoic fish evolution. He held postdoctoral positions at the Australian National University (1984–1985, Rothmans Fellow), the University of Western Australia (1986–1987, Queen Elizabeth II Award) and the University of Tasmania (1988–1989, Austrailian Reasearch Council (ARC) Fellow), before taking up a position as Curator in Vertebrate Paleontology at the Western Australian Museum (1989–2004) and then as Head of Sciences at Museum Victoria (2004–2009).

Adrian Norris: Chair, International Council of Museums, Working Group on the Art of Taxidermy and Its Cultural Heritage Importance, Leeds, UK. Adrian Norris was for many years the Senior Keeper of Natural History with Leeds City Museums in the UK. Since retiring from this position, he has become involved with the International Council of Museums Natural History Committee as co-chair of the Working Group on the Art of Taxidermy and its Cultural Heritage Importance, as well as on their Ethics Committee. Within Yorkshire, he also chairs the Yorkshire Naturalists' Union's Natural Sciences Committee and is on the executive of the Yorkshire and Humberside Biodiversity Forum. Prompted by the ICOM NATHIST meeting held in Wellington in 2006, Adrian Norris contributes to the repatriation debate by raising issues often overlooked by New Zealand's current focus on human remains. He has been the recorder for mollusca for the Yorkshire Conchological Society and the Yorkshire Naturalists' Union for over 25 years and is a Life Member and a Past President of the Society.

Ulrike Stottrop: Deputy Director, Collections and Science and Director Mineral Museum, Essen, Germany. In 1984 Ulrike Stottrop began as a geologist at Ruhrland Museum and became Head of the Geological Department and Head of the Mineral–Museum Essen, which is part of the Geology Department of the Ruhr Museum. From 1986 to 2007, she was Associate Director of the Ruhrland Museum. In 2008 the Ruhrland

Museum became the Stiftung Ruhr Museum, and since that time, she has been the Head of the Science and Collections Department. She has realized more that 50 exhibitions, many with the thematic focal points: the man-made nature of the "*Ruhrgebiet*" (Ruhr Area), the cultural history of nature and natural science and art, as a natural scientist working on mutual projects with artists, historians and cultural scientists. Between 2004 and 2008, she was Speaker of the German Natural History Museums in the German Museums Association and has served on the Board of ICOM NATHIST as a member and treasurer.

B. Venugopal: Director, National Museum of Natural History, New Delhi, India (Government of India, Ministry of Environment and Forests). Dr B. Venugopal is at present Director of the National Museum of Natural History (NMNH), New Delhi, India. Before assuming the Directorship, he has been involved in the establishment of two regional museums of Natural History at Mysore and Bhopal in India. He has also worked as Registrar of National Museum Institute, a Deemed University for Museology in India. Dr. Venugopal has been trained in Museology from NMNH, India (New Delhi, 1985) and Leicester University, UK (1991). He has written a chapter for the book "Museums, Media and Message" edited by Dr Eilean Hooper-Greenhill and published by Routledge UK. He is actively involved in the museum movement at the national and international levels and organized a large number of professional programs in museology and museography in India. This included the First South Asia Museums Workshop (Bhopal, 2000), ICOM Committee for Education and Cultural Action (CECA) First Asian Workshop on Museum Visitors (New Delhi, 2002) and the first National Conference on Intangible Natural Heritage (Calicut, 2007). He has participated in a few ICOM programs on Museums and Intangible Heritage (Shanghai, 2002, Seoul, 2004 and Hyderabad 2008). At the international level, he has been associated with ICOM. He was an Executive Board Member of ICOM-CECA for two terms. At present he is an Executive Board Member of ICOM NATHIST.

Alan L. Yen: Research Leader, Invertebrate Sciences, Biosciences Research Division, Department of Primary Industries, Australia. Dr Alan Yen is the Statewide Leader for Invertebrate Sciences at Primary Industries Research Victoria, Department of Primary Industries. He is a graduate of La Trobe University and held positions at Monash University, Museum of Victoria and the Arthur Rylah Institute. His research interests are insect-Eucalyptus and insect-Acacia interactions, invertebrate surveys, invertebrate conservation, ethnoentomology and now conservation biological control.

Index

Page references in *italics* refer to photographs, illustrations or tables.
Page references in **bold** refer to the main contribution from each author.